**Terrence Hedley Burchell Butcher**

# My Life
# My Story

## THOUGHTFUL RECOLLECTIONS FROM AN INNOVATIVE MIND

COPYRIGHT © Terrence Hedley Burchell Butcher 2024

All rights reserved. No part of this book may be reproduced by any mechanical photographic, or electronic process, or in the form of a phonographic recording: nor may it be stored in a retrieval system, transmitted, or otherwise be copied for public or private use other than for "fair use" as brief quotations embodied in articles and reviews-without prior written permission of the publisher.

DISCLAIMER

This is a work of creative non-fiction. All the events in this memoir are true to the best of the author's memory. The author disclaims any liability for any information or commentary that proves to be inaccurate or incomplete.

---

Ghostwriter : www.yourmemoirpublished.com
Editor : www.poppyspagesediting.com
Cover Design & Interior Format: www.inawonderworld.com
Front Cover Photo: John Pforr, Bruny Island, Tasmania.

Paperback ISBN: 978-0-6454756-6-1
Hardback ISBN: 978-0-6454756-5-4
Ebook ISBN: 978-06454756-7-8

1st Edition September 2024

I wish to dedicate this book to
my lifelong friend,
Ian Sydney Hamilton.

To all my friends and relations,
please don't read my book looking for
a reference to yourself in it.
It won't be there as it is just a book about me.

# CONTENTS

| | |
|---|---|
| **Prologue** | 9 |
| **Chapter 1 – 35,000 Books** | 15 |
| **Chapter 2 – Leaders Are Born** | 19 |
| **Chapter 3 – The Butcher Ancestry** | 27 |
| John Hunt Butcher (1781-1839) | 27 |
| Sarah's brother, William John Burchell (1781-1863) | 29 |
| Edward William Burchell Butcher (1829-1895) | 38 |
| Edward William Norton Butcher (1854-1918) | 42 |
| Reginald Charles Burchell Butcher (1902-1987) | 46 |
| **Chapter 4 – My Life Story Begins** | 53 |
| Shubert's Marche Militaire | 55 |
| 1st Mosman Scout Group | 59 |
| The Montsalvat Artist Colony | 66 |
| World War II | 71 |
| Western Australia | 74 |
| Erecting Camouflage Tree, Menangle (1944) | 79 |
| The War Ends; Back to Sydney (1946) | 85 |
| Prouds the Jewellers (1948) | 90 |
| No longer an only child | 95 |
| **Chapter 5 – Engineering Solid Foundations** | 101 |
| Fitter & Turner Apprenticeship (1948-1952) | 101 |
| Royal Australian Navy Reserve (1949-1955) | 112 |
| Shell Oil Company (1953-1954) | 121 |

## Chapter 6 – World Adventures (1955-1961) — 131
Voyage to England — 131
British Broadcasting Corporation (BBC) — 144
High Definition Films — 147
My first film production on 35mm — 159
Jacquie — 162
Voyage to Canada (1956) — 169
Canadian Broadcasting Company (CBC) — 172
My son, Rolf (1957) — 174
My daughters, Sasha (1963) and Belinda (1964) — 177
Dominion Electric Company — 187

## Chapter 7 – Australia (1961) — 191
Gollin & Company, the vario-klischograph — 193
Sydney University (1961) — 195
1st Restoration: 120 Jersey Road, Woolahra (1962) — 201
Laurence & Allumask — 209

## Chapter 8 – TEB Plastics Enterprises (1965-1973) — 211
2nd Restoration: 122 Hopetoun Avenue, Vaucluse. — 214

## Chapter 9 – New Beginnings (1975) — 227
Marian — 227
3rd Restoration: 215 Cleveland Street, Redfern — 230
Palingenesis Antique Restorations — 241
4th Restoration: 56 Pitt Street, Redfern — 242
Black Tie 50th Party — 250

## Chapter 10 – My Collections — 259
   Meccano — 262
   Losing my early collections — 267
   My Collection List — 268
   Envelopes and Transcripts (circa 1800) — 270
   Seeds from William Burchell (circa 1800) — 271
   Sydney Harbour Bridge Collection — 272

## Chapter 11 – Collecting and Restoring Old Tools — 281
   Tools are an extension of one's hand — 281

## Chapter 12 – My Inventions — 293
   Apprentice tool making — 294
   CSR Development of Masonite & The Aerosol Canister — 294
   Serrated Glad-wrap Cutter — 294
   Lindemans Champagne Cork Remover — 295
   Car Cigarette Dispenser — 296
   Hand Surgical Instruments — 297
   Nasal Continuous Positive Airway Pressure Mask (1980) — 298
   Tow Bar Ball Cover — 299
   Smiley - The Original Emoji (1971) — 301
   Plastic Clamp Toy Rings — 304

## Chapter 13 – Bruny Island (1996-2020) — 307
   First Night in Redfern — 320
   The Bruny Island Men's Shed — 339
   The Bruny Lifestyle — 345

| | |
|---|---:|
| Love of Music | 355 |
| Special Friendships | 358 |
| **Chapter 14 – Sunshine Coast (2020-2024)** | **365** |
| Thirty-six abodes | 369 |
| Climbing the Bridge | 372 |
| **Epilogue** | **375** |
| **Acknowledgement** | **377** |
| **Bibliography** | **381** |

# PROLOGUE

I was at one of those parties when you sort of helped yourself to drinks – there were plenty of snacks and dips on the table – when I got into a conversation with a chap about the same age, in his sixties, discussing politics. As our exchange came slowly to an end, I casually asked him what he did in life.

He replied, *"Accountancy has been my whole life. I excelled in mathematics at school and took a job straight away. After getting my leaving certificate I spent my whole life with a bank at their head office in the city and have only recently retired, after forty-seven years. It was interesting, but my favourite pastime has been growing roses. Oh yes, I had two children and two wives. But both of my wives have died and I don't know where the kids have gone. I'm thinking about writing a book on propagating roses. By the way, what did you do?"*

I stammered a little, then answered, *"At school, I got as far as the intermediate certificate and I too excelled at mathematics. I was in the school tennis and swimming groups, also out of school. I liked to use my own pocket money, as there was little coming from my parents, especially during the WWII period. I joined a bakery delivering bread on Saturday mornings from the bread cart, which was being pulled along by an old retired racehorse. I caught crabs in the Swan River and after cooking the crabs, I sold them to the ferry passengers. We also sold papers on the street corner by our local pub until my dad discovered my little enterprise and forfeited this, as he was the managing director of the newspaper.*

*In my last year of school I worked at Prouds the Jewellers shop selling pens, cigarette lighters and gold cigar cases until it was time to join an engineering firm to start my career serving a five-year apprenticeship as a fitter and turner.*

*As the Korean War was heating up, I enlisted in the Australian Navy Reserves for one night a week and occasional weekends. I had training on warships and in submarines, rising from Seaman to Stocker then Engine Room Artificer $5^{th}$ class to Engine Room Artificer $4^{th}$ class. Eventually, I was selected for Officer Training, but at this point, with my apprenticeship completed, I joined Shell*

*Oil Company and worked for a year in their film department before making it overseas. My first job in England was with a film company that made equipment to produce films for television cameras.*

*At this point, having met the girl of my dreams and married, we both decided to live in Canada where I worked for the CBC, the largest broadcasting company in the country. My role was to maintain all the shop floor equipment. Out of town, I was offered a job to run their production line with a much higher pay rate, but after a year or so (along with the birth of my son), we returned home. My wife being a Kiwi, we had to stop off in New Zealand where her father offered me a job working in his brick factory. I turned him down and this upset my wife considerably, but we returned to Australia.*

*Quite quickly we set our sights on buying a terrace house and turned it into a boarding house, to help pay it off. Although I did get a job selling printing machinery, before long I quit that and took a job on the staff of Sydney University. It was to build the world's largest radio telescope. There were just three of us to get this project started. When it was completed, some four years later, and having learnt about plastics, I started in my garage at home, making small plastic parts for the industry. Before long I became a managing director of my own company and moved into larger premises in the city. This is when I introduced Smiley to Australia – and maybe the world.*

*But a disaster struck my company because I didn't get government help as promised and went out of business. During this time we had two more children and then divorced. My wife took off with the family and moved to Queensland. So turning my hand back to plastics, I got a job with a small firm, met a lovely girl, and married her.*

*On meeting an antique dealer, I was asked if I could make various items to restore his antiques that he was looking to re-sell. I soon opened my own company making pieces for most of the dealers that lived and worked in Sydney.*

*On reaching my retirement years my wife, who remained childless, decided that we should go and live in Tasmania. This was a great adventure, where I started a Men's Shed and was the president for the time we were there. However, my wife died, and I moved on to Queensland to be close to my family once more. Then I decided to write this book; no, I didn't go in for roses, but I became a successful collector."*

To this day, I still don't know exactly what it was, or how I could say, that I earned my pay.

I feel very fortunate that a lot of interesting things have happened in my life to write about that fill my memoirs. Why don't we embark on this journey together?

# CHAPTER 1

## 35,000 Books

I had a dream this morning, waking up thinking about my memoir, and I've decided to call my first chapter, '35,000 Books'. They have meant so much to me throughout my life. Books, yes books, are one of mankind's greatest inventions.

Why, you may ask? Just look at what they contain: human history, things to learn, things to do or how to make things. Pick any subject and there is a book already written about it. At one time my wife and I had over 35,000 books in my collection; until I decided to leave Tasmania and move to Queensland.

My first book I couldn't read (I was just too young), but my father read to me the wonderful stories by A. A. Milne, who wrote all the Winnie-the-Pooh stories. My dad chose a voice to put each character in and if there was one of Ernest Shepard's drawings, he would show me the picture and describe it. I would sit beside him as he read, and soon I was able to identify the written shapes into words. Eventually, I was able to read the stories for myself and so began a long life of reading and collecting books. I still have the Pooh Bear books dated 1933.

Much later in life, when I was living near Buderim in Queensland, author and artist Richard Timperley's wife, Hazel, and daughter, Georgie, lived in the area. By this time Richard had passed away, but not before publishing a book called 'Heroes Galore', which debates much of my family history. I gave my copy of the book to Dick Geeves, a friend in Tasmania, to read. A couple of months later he came back to me with his tail between his legs. *"Terribly sorry, I lost your book. I must have left it on the bus to Hobart, but I've found a copy on Google and it is coming from New Zealand."* He had acquired a precious copy, not only signed by Richard but a gift to his wife. This book is now in the State Library of Western Australia.

The book 'Ups and Downs' reveals how the Butcher family donated the second plane that formed the Australian Air Force. The book also covers how the Johnston family were related to the Butcher family. My mother often referred to Freddy Johnston, the Surveyor General, who was responsible for measuring up the railway lines from Adelaide to Perth, the Nullarbor Plains and so on. The book's writer, Wendye E. Camier (nee Johnston) is my cousin. Her father was Edgar Charles Johnston (1896-1988) and her grandfather was Harry Frederick Johnston (1853-1915), The Surveyor General of Western Australia, as noted on page 50 of her book.

# CHAPTER 2

## Leaders Are Born

I picked up a card the other day displaying the meaning of my name, 'Terry' (short for 'Terrence', an Irish name from my Irish mother). "You are a born leader," it said, as if by sheer destiny. The leadership quality has been and still is something of great importance for me – to act with full autonomy in all my work and life choices. A leader feels independent and takes many risks whereas a follower is happy to sit in a safer position in life.

My earliest memory was when I started my first 'gang' as a child and wanted to be the leader. Later I joined the scouts and needed to become the scout leader. As an adult, during my working life, I was never fully satisfied until I was the top man.

My innovative mind could not cope with being in the background. Whenever I sat in a meeting or found myself in a place not leading the subject matter, I would feel completely jealous of the person in charge. I was not happy in life until I became a leader in my many fields of expertise.

Being a leader does not always mean I want to be in control of others by being the boss. I simply enjoy having control over events and being autonomous to move forward on my ideas or goals at hand. Many times, I have been placed in managerial or supervisory positions. But when I established my own company in Sydney, I was immensely satisfied. I had created something very special. I had people working for me and yes, I was their boss, and simply loved it.

I take great pride in my level of self-sufficiency and creativity. It feels natural for me to step forward, in front of any crowd, to lead a project. Following other people's rules or regulations often annoys me. I prefer to set my own rules for people to follow my creative thinking. Although in any disagreements, I never became a bully. I was most happy to discuss other people's ideas to find amicable solutions. Many people have come to me for advice and run their ideas across me first. I love brainstorming these creative endeavours, ensuring practical steps to see projects through to fruition.

I started an antique tool club, The Trades Tool Group in Sydney, which is now the biggest tool club in Australia. Our first meetup hall was at the local mortuary. The club was renamed in 1995, becoming The Traditional Tools Group. On Bruny Island in 2020, a group of us got together to start a Men's Shed and I was elected as president. The Men's Shed is still going strong.

I never liked to settle for mediocrity, I always pushed every project to reach its full potential. My ultimate aim was perfection. Generally, I never considered it was an ego thing – "Look at me and see what I can do" – it was about the situation, to create something wonderful. This has been my driving force in life. It makes me different from many other people, who don't understand my need to strive for the best. After the success of many goals in my life, a welcome ego boost now and then didn't go astray for my well-being. I had a lot of people to help me in my company and was a Chairman of the Board. Being a director may sound impressive, but it was only for legal matters that I had that title.

My father may well have had similar ambitions in life, but I never saw him working at things specifically. He was a quiet achiever; he plodded through until he could improve himself on what he wanted to do. Dad loved painting, which became purely a solo performance. He wasn't interested in having exhibitions to show the world his work. His paintings were for his creative satisfaction. The same can be said about his love of pottery. However, to survive, he had to make a lot of fairly common ceramic pots, for example, salt pepper pots or tobacco pots. Both pottery and painting were tremendous therapy. He loved sitting at the potter's wheel. Parallels can be drawn to my dad's achievements though. He made a modest living from selling pottery whereby I created a career out of making things with plastic.

I also started Palingenesis (Latin for rebirth), a company that restored furniture. Here I became a well-known leader in Sydney for my restoration business only because of the pure quantity of my workload. I was the only person working in this field with full time hours. Many others only worked part time or as their hobby restoring furniture. I made a very successful full time job out of it. I was highly ranked in the trade and that boosted my ego tremendously.

During the mid-1930s, when my family moved to a new house in Mosman, I befriended a boy next door, a bit younger than I was, maybe five years old. I took on the responsibility of looking after him and together we often ventured from home, exploring an area called Reid Park in Mosman Bay. I was constantly criticised by his mother for leading him astray. My mother was not too happy about that either. One adventure that got me into the most trouble was the day we explored the waterways nearby. On one side, there was a drain that ran underneath the park from the creek, and on the other side was a water channel, which took care of the rain during the wet season. It was fun to go into the channel, find the outlet and crawl the entire length of the pipe, coming out among the trees (which under normal weather conditions, was a small creek). When our mothers found out where we had been playing that day, I was reprimanded. This adventure went to show that my leadership wasn't always for the best; sometimes I led myself and others into trouble.

I joined the Mosman Cubs, which being part of the scout group held their meetings in the barn at Mosman Bay. It was fun on Monday nights when we all met up. Most of our training came from the leader of our little patrol group. All the groups consisted of about five or six members. Our leader had been a cub for some years. He had been shown how it was done and passed this information to his little patrol group. We did have group meetings where all the cubs came together as a pack and were taught by the pack leader, an adult known as Akela. I was a quick learner and couldn't wait to become a Patrol Leader. Unfortunately, the war came and we moved to Western Australia before I had the opportunity.

Living in the suburbs of South Perth in Western Australia, I gathered many local kids. We would go bicycle riding together or swimming in the Swan River. My mother gave me an old sheet on which I printed the letters PX GANG then hung it on a tree in our backyard where I loved getting everyone together to plan our events. Saturday night was always fun. We would all go out on our bicycles to a local outdoor movie theatre to form our cheer squad.

We returned to Mosman when the war was over and the first thing I did was to rejoin the Boy Scouts club of the 1st Mosman, which incidentally was the first Troup in Australia. I was put in a group where all the cubs were soon to become scouts. I was a bit older and the Patrol Leader made me his second in command. He notified me that soon he was going to join the Rovers and I would have to go to the training centre at Pennant Hills to learn how to be a Patrol Leader. It was a lot of fun. There were scout groups from all over the northern suburbs. Eventually, I became a Patrol Leader and though we weren't the top patrol group I'm sure we had the best time. After some years I yearned to take on the job of Troop Leader, a responsible position. The Troop Leader would set up his group ready for inspection by the Scout Master, to check uniforms and shoes were effectively perfect, quickly chastising those that stuffed up. I joined the Rover Crew. We had a very good Scout Master who was also our Rover Crew Leader.

# CHAPTER 3

# The Butcher Ancestry

My family heritage is a testament to the strength and courage of pioneering individuals, who shaped the life I have enjoyed.

### John Hunt Butcher (1781-1839)

My great, great grandfather, John Hunt Butcher, was born in Hascombe, Surrey, England. I donated an oil painting of John Butcher (290 x 235mm) to the Tasmanian State Library in Hobart. I saw the sale document but no year was shown. John married Sarah Burchell (1791-1872) in Westminster, England in 1808. Sarah came from a family of nine children. Her parents were Mathew Burchell and Jane Cobb. Sarah named most of her children 'Burchell Butcher'. It is believed that John, a Justice of the Peace and Police Magistrate, was appointed by

King William IV to set sail for the Colonies. John and Sarah's tombstones lie at St Davids Park Cemetery, in Tasmania however all the tombstones have now been removed to build a wall around the park cemetery.

John Hunt Butcher (circa 1830).

# Sarah's brother, William John Burchell (1781-1863)

William and Sarah's father owned the Fulham Nursery and Botanical Gardens in England. William became a well-known British explorer and botanist who enjoyed an adventurous life. Firstly, he travelled to St Helena, a British outpost in the Atlantic Ocean. It was here he met the exiled Napoleon Bonaparte (1769-1821). Later, in 1810, William travelled to South Africa to study the fauna and flora. William had a talent for drawing and would sketch with detailed precision, documenting his travels and discoveries. By the time Sarah was living in Tasmania, she had started receiving plant specimens from her brother. William sent cuttings and seeds to grow in her herbarium. I had volumes of William's notes and passed them to my son, Rolf.

William John Burchell (circa 1854).

In 2015, Penguin Books published 'Burchell's Travels', being the life, art and journeys of William. This book was extensively researched and compiled by Susan Buchanan.

William had discovered a unique zebra in Africa, which led him to name the zebra after himself: 'Burchell's zebra'. Each zebra has unique stripes enabling identification; Burchell's fade out on the legs. William's early findings are now being fully recognised. His drawings and discoveries have been showcased on African stamps to celebrate many anniversaries of his life's work. There are exhibits at Oxford University, in England, side by side with Charles Darwin, of his extensive collection of artefacts from his South African discoveries. If I had a dinner party of any six people, Dr Burchell would be on the invitation list. He was a naturalist and I would find it immensely interesting to discuss his findings.

Burchell's Zebra and Terry at Taronga Zoo, Sydney (circa 1940).

## *Van Diemen's Land*

Van Diemen's Land was a British Colony settled between 1803 and 1830 at which time it joined five other colonies to form the Commonwealth of Australia. In 1856 it was renamed Tasmania.

John, Sarah and their five children left England aboard the *Deveron* arriving in Van Diemen's Land on the 19th June 1822. They brought with them a herd of Saxon Merino sheep on a three-month voyage. It was announced that all the sheep travelled well. The Hobart Newspaper reported the family's arrival and claimed their bringing of the first Merino sheep into Tasmania.

## *Richmond*

John and Sarah were originally given a grant of land located in Hamilton. John then purchased 1800 acres of land at Richmond in 1828 from Lieutenant-Governor William Sorell and named it 'Lowlands'. There are conflicting stories about what the transactions were between the government and John Butcher, but a hundred acres of John's land was eventually acquired so that the township of Richmond could be built. The Lowlands property was rich in siliceous sandstone and most of the freestone houses in Richmond came from Butcher's Hills.

John and Sarah settled in the Hobart township where they stayed while their house was being built on their property. Their land was thirty miles west of Hobart. They built a stone house naming it 'Lowlands' in Richmond, which stayed in our family until the early 1850s. The Bill of Sale document is in my collection today. By then John and Sarah had seven children: Harriett, Jane, Martha, John, Sara, Caroline and Edward.

The building of the Richmond Bridge, now the oldest bridge in Australia, was recommended by the British Royal Commissioner, John Thomas Big, when he visited Van Diemen's Land in 1820. On the 13th December, 1823, the Hobart Town Gazette and Van Diemen's Land Advertiser announced, *"The first stone of the bridge over the Coal River was laid..."* The bridge was built by convict labour and the stones were quarried from Butcher's Hills (Lowlands quarry). Legend has it that the men were required to haul the stone by hand cart, two pulling and one pushing. It is said that a particularly brutal overseer rode on the top of the load when fancy suited him. The men attacked this man one misty morning and threw his broken body onto the rocks below the bridge. This bridge is still in operation today. It opened up access between Hobart and Port Arthur. There is a plaque on the bridge referring to the construction, 'using locally quarried freestone'. The quarry on Lowlands was often a hiding place for bush rangers, stealing silver and turning it into bullets. Many buildings in both Richmond and Hobart used this stone. The quarry was quite profitable and contributed significantly to the development of Richmond.

There was already a Catholic Church in Richmond, however the Church of England wanted to buy land to build their own church. The second church built in Richmond was that of the establishment: St Luke's Church of England in Torrens Street. The foundation stone was laid on the 3rd February 1934 by Lieutenant-Governor Sir George Arthur (1784-1854). He governed from 1825 until 1836, the longest serving Colonial Governor. Again, the church was built with the assistance of the convicts, using sandstone from the Lowlands quarry. The land on which the church stands was exchanged by John Hunt Butcher for a grant of 640 acres of bushland further upriver. There is now a plaque in the church garden referring to the sandstone used to build that was bought from John.

In the church there was a private door used by the Butcher family and some parishioners, however convicts always came through the main door. Pews at the front of the church were for the privileged to sit in. Prisoners always stood in the back. John Butcher died in 1839, not long after his sons left the property.

The northern section of the Richmond Gaol was built in 1825, five years before Port Arthur. The gaol was not only used for convicts but also for free people who broke the law. As John was a Police Magistrate, he was often called upon. In 1825, the Colonial Times reported, *"We are sorry to state, that a daring robbery was committed last week, at the house of Mr Butcher, at the Coal River, which was entered by ten armed men with their faces blackened, who plundered the house of everything valuable."* The bushrangers appear to have been Brady and McCabe. During this era, Tasmania's most notorious and celebrated bushrangers were elusive to the authorities. They were in search of silver that could be made into bullets. Thankfully, in Butcher's home, his silver teapot was well hidden behind the lounge. I have since had a replica made of this teapot, using the same silver makers, and it is in my collection today.

The self-proclaimed King of Iceland and world-renowned explorer, Jurgen Jorgensen (1780-1841), challenged one of Butcher's sons to take a wagon from him. He was charged. The court ruling papers came through an auction house and

I wanted to buy them. The government insisted they go to the State Library Archives, however I kept copies.

### *Butcher's Hill Vintage*

The Coal Valley region had very fertile soils. Today that area is mostly grape growing country. During the 1820s, Richmond was only a small settlement, comprising a baker, a miller and possibly a hotel. A large successful winery is now located in this area; Pooley Wines was established in 1985. By 2007 they had acquired a second vineyard producing their first vintage, called 'Butcher's Hill'.

# Edward William Burchell Butcher (1829-1895)

My great-grandfather, Edward William Burchell Butcher, was only twelve when his father died. He and his siblings were left with inheritance monies to finance their education. Edward returned to England and studied at Kings College in London. He later married Maria Susan Shaw from Jamaica. Susan's father, Major Charles Shaw, was a Police Magistrate. Together they had five children: Frances Emma (1838-1846), Edward William Norton (1854-1913), Maria Louisa (1856-1886), William James Burchell (1858-1944) and Charles John Hunt (1860-1931).

*Exerts from newspapers [E.W.B. Butcher]*

The Tasmanian Colonist, 18 September 1851 *"Maginnis v Crooke. The following gentlemen were sworn as Jury: Joshua Penny (foreman) John Dixon Lock, Edward William Burchell Butcher..."*

Colonial Times, Hobart, 18th October 1855 *"William Currey, free by servitude, charged with stealing on the 30th July last, six bushels of barley value 3 and other articles, the property of Edward William Butcher. Edward W. Butcher sworn - resides in Richmond. Knows the prisoner. He had been employed by the witness in thrashing grain."*

The Inquirer and Commercial News, Perth, WA, 20th December 1889 *"New Justices of the Peace for this Colony - James Gordon Knight, Edward William Butcher, Daniel Matheson, Elias Solomon ..."*

The Mercury, Hobart, 31st January 1895 (under Deaths) *"BUTCHER, On January 28, 1895, at Carnarvon, W.A. Edward William Burchell Butcher, late of 'Lowlands', Richmond Tasmania, aged 68 years."*

The West Australian, 14 February 1895 *"The late Mr E W Butcher (from a correspondent), Mr Edward William Butcher, at the age of sixty-seven, breathed his last at the Mission House, Carnarvon, the residence of his daughter, Miss Butcher, on the evening of Monday, January 28th, and was buried in the Carnarvon Cemetery on Wednesday, January 30th. He passed away peacefully in the chair, without struggle, the mainspring of life having quietly snapped through heart failure. A gentleman he was whom to know was to appreciate. The estimation he was held in was demonstrated by the fact that all the business of the town was suspended the afternoon of the funeral; all flags were half-masted. Every man, high or low, united in offering the last respect possible by following to their last resting place the remains of the deceased. The chief mourners were Mr James Butcher, Mr Charles Butcher and Mr C.D.V. Foes, our respected Resident Magistrate. The pallbearers were Messrs. J. Brocman, G. Baston, J. Mansfield and J. Pincombe, wearing the regalia of the Masonic Order, of which Order the deceased was a brother. The funeral service was read by Mr. John Brockman, in accordance with a long previously expressed by Mr. Butcher. All I can say of the deceased is that he was a gentleman in the full sense of the term.* **He had partaken to the full of all the varied vicissitudes occurring in the life of an adventurous speculative Australian, at times riding gaily on top of the wheel of fortune, and sometimes with the wheel rolling over him, but never crashing him, owing to his indomitable spirit.** *He was a true type of our British pluck.*

*Mr. E.W. Butcher was born on the 12th of May 1827. He was the youngest son of Mr John Hunt Butcher of 'Lowlands' Richmond, Tasmania and was educated at King's College, England. When as a young man he tried his fortune on the Bendigo diggins, with a fair amount of success. He then managed for some time East Loddon Station. Afterwards, for many years he owned Towangay Station near Wickliffe, Victoria. He came to West Australia about 1876, and took up and stocked Monroe and Berringana Stations on the Murchison River. From squatting he turned his attention for some years to pearling in Shark Bay. He leaves behind to lament his loss, his wife, Mrs Butcher, three sons (Mr Norton Butcher, Chief Draughtsman, Tasmania, Mr*

*James and Mr Charles Butcher of Boolithana, Carnarvon) and two daughters \*\*Mrs H Johnston and Miss Butcher. \*\*Maria Louise Butcher (1856-1921) married 1879 (Geraldton) Harry Frederick Johnston (1853-1915). Harry F. Johnston was the Surveyor General of Western Australia for 20 years. Children: Edward Betram 1880, Henry William 1881, Hubert Cookburn 1883, Frederick Marshall 1886, Hilda May 1887, Sidney St. Maur 1888, Dorothy Renee 1890, Gertrude Frances 1891, Rose Ethel 1893, Edgar Charles 1896, Alfred Laurence 1898."*

# Edward William Norton Butcher (1854-1918)

My grandfather, Edward William Norton Butcher, was born in Tasmania in 1854. He wooed the eldest daughter of the Timperley family, Maud Elizabeth (1860-1922). On the 14th January, 1882 in Geraldton, Western Australia, Edward and Maud were married. Together they sailed to Melbourne and eventually over to Tasmania in 1882 where Edward took up a permanent position as a draughtsman-surveyor in Hobart. They had nine children: Ethel Maud (1882), Edward William Norton (1883), Dora Cockburn (1884), Ernest Norton Shaw (1885), Francis John Hunt (1888), Clarice Helena (1891), Marjorie Constance (1892), Herbert Bradley Cockburn (1895), Reginald Charles Burchell (1902).

Edward William Norton Butcher.

- The first child born was my Aunty Ethel Maud (1882-1913). She was born in Bourke, on the Darling River in New South Wales. By mid 1882 they were back living in Tasmania. Ethel married William Frederick George Collett. She died at childbirth being survived by her son Arthur Frederick Collett (1913). Arthur had two children: Harvey Peter and Katie Harriet.
- The second child, my Uncle Edward William Norton (1883), married Muriel Leura Badham of York and Katanning, Western Australia. My Aunty Muriel was affectionately known as 'Moosie'. They had two sons: Alan and Raymond.
- The third child, Dora Cockburn (1884) and the fourth child, Ernest Norton Shaw (1885), lived only for a brief time, possibly less than a year or two each.
- The fifth child, Uncle Francis John Hunt (1888-1892) lived for three or four years.
- The sixth, Aunty Clarice Helena (1891) married Bertram Henderson and had one daughter, Barbara.
- The seventh child, Aunty Marjorie Constance (1892) married Harry Angove and had two children; William and Gwenda.
- The eighth child, Uncle Herbert Bradley Cockburn

(1895-1896), lived less than twelve months.
- Then along came my father, being the ninth child of the clan, Reginald Charles Burchell (1902-1987)

# Reginald Charles Burchell Butcher (1902-1987)

My grandmother gave birth to her last baby at aged 42 years, my father, Reginald, who later came to be known as 'Butch'. I would imagine that for his sisters; ten-year-old Clarice and 11-year-old Marjorie, the arrival of a baby brother was extremely exciting. Whereas Reginald's surviving older siblings, Dora, Edward and Ernest, were now growing teenagers forging their own lives. The family's home was called 'Buena', situated in the suburb of Bellerive in Hobart.

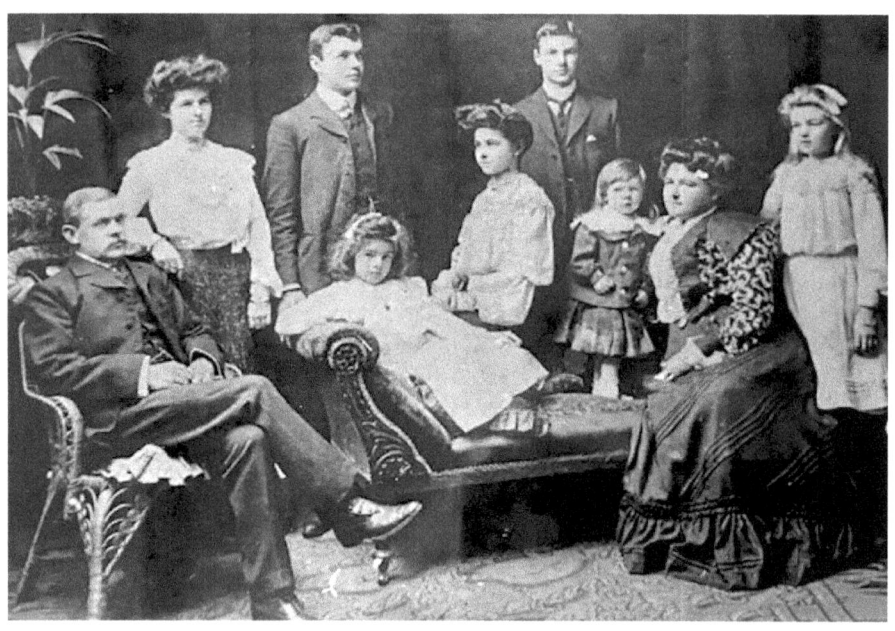

The Butcher family portrait,
Butch, the youngest child, 3rd from right.

From time-to-time, Uncle Edward took his now seven children (two had died very early in their lives) to visit their relatives at the Lowlands property in Richmond. Aunty Dora recalled that at one time, Lowlands was leased to the Jacobs family, descendants of the original servants the Butchers had brought out from England. My father spent a lot of time at the Lowland's farm. The home was burnt down circa 1910 but was rebuilt. I revisited Lowlands after another fire had gone through the property. I was able to square up the engraving on the fireplace, a part of the house that remained undamaged. In 2010, when I was living in Cloudy Bay, my next-door neighbour happened to be a relative of the Lazenby family, who were living in Lowlands at that time.

My grandparents, Edward and Maud, travelled by sea to Western Australia from Hobart with their families. My father, Butch, was about twelve years old when he arrived in Perth. Edward and Maud became divorced in Perth. My grandmother, Maud, died in Albany in 1922.

Edward went on an exploration throughout Western Australia with my great-grandfather, Edward William Burchell Butcher. They became business partners in Broome and collected pearls by employing divers from Japan.

Butch went to a boarding school at Guildford Grammar in Perth. After finishing school, he went droving with his uncle Jim, William James Burchell Butcher (1858-1944), whom he loved. Jim's business partners were his brother Charles and brother-in-law John Brockman. Together they owned a large property at Boolathana in the Gascoyne region of northwestern, Western Australia, grazing cattle. Jim was married to Margaret Harriet 'Maggie' Brockman. By 1901, Jim and his family resettled from the isolated station of Boolathana to Garden Hill, Perth, and later became a Senator for the Western Australian Parliament. He died in 1944 at Nedlands and is now laid to rest in the Karrakatta Cemetery. On the death notice of my great uncle Jim, it was noted that the Butcher Family had donated a plane to the Royal Australian Air Force, then called the Australian Flying Corps, during WWI. This plane was named 'Boolathana'.

Robert Cecil Jones (1864-1951) married my aunty, Frances Emma Butcher (1862-1946). The Jones family owned a prosperous farm at Hampton Hill in the Bulong region, northwest of Western Australia. Here my father enjoyed working whilst gaining an informal education during his teenage years. Becoming a young man, Butch thought perhaps he should give up the droving and farm work to find a proper job.

My dad walked into the Perth Daily News office and simply asked for a job. It was agreed that he could start with the company as a copy-boy but only on a trial basis. He was never officially told his trial was over and remained with the company for fifty years. Dad loved the newspaper world and was soon elevated to a journalist position covering many events, including sports reporting. It was during this time he met a lovely Irish girl from Belfast. Her name was Rita, short for Margarite Anna Victoria Harris (born 30th December,1900). They eventually got married, circa 1927 (being unable to confirm the date, as I only have an extract of my birth certificate). My mother was working at this time but I do not recall what her job was. My parents were not well off and rented an old fisherman's cottage: number five, Hardy Street, South Perth, on the shores of the Swan River.

Reginald Charles Burchell Butcher (circa 1930).

## *The Charles Kingsford Smith Connection*

Charles Hedley McNish worked for the Wakefield Oil Company. This company had the privilege of supplying oils and greases to aviation pioneer Charles Kingsford Smith (1897-1935) at his Western Australia Aerial School. Charles Kingsford Smith had recently completed the first transpacific flight from the United States of America in 1928. It was a three-leg journey, from California to Brisbane via Hawaii and Fiji. Charles Ulm (1898-1934) was his co-pilot. All three Charles lived in South Perth at this time and became good friends with my parents, often visiting for family meals. Charles Kingsford Smith and my mother became tennis partners. I have a framed photograph of them all visiting our family, standing in the backyard, taken before I was born.

When I was older, I remember meeting Charles Kingsford Smith and Charles Ulm. Within my collection, I have a book signed by them, highlighting their first nonstop flight from New Zealand to Australia, which was achieved in the same year as the transpacific record. I also have in my possession the 78 voice recordings of Charles Kingsford Smith's first flight from the United States of America to Australia, which were given to my parents in South Perth after the event. My collection also contains celebratory postage stamps on envelopes signed by Charles Ulm commemorating the first New Zealand to Australia flight.

My father was intelligent and had the good fortune of being educated at a Grammar School. His closest friend was Charles Hedley McNish who became my godfather. That is how I was given 'Hedley' as my middle name.

Charles Kingsford Smith, Charles McNish (My godfather-to-be), Dad, Mum (pregnant with me), Mrs Ulm and Charles Ulm, 1931.

# CHAPTER 4

# My Life Story Begins

I was born at home, in our tiny fisherman's cottage, 5 Hardy Street, South Perth, on Saturday the 31st October 1931. Our home was modest and the nursery was much the size of a storeroom. To this day I remember the nursery room and our backyard. We had regular kookaburra visitors who would perch themselves on our fence. I was a healthy boy with brown eyes and very dark black hair. Bobby Gaffney was the midwife for my birth and became very good friends with my mother. They remained friends most of her life until she died in NSW sometime around 1970.

My mother and I often had picnics on the beach, just outside our house on the Swan River. We had bread, butter, jam, and a breadboard with a bread saw. One day, as she slept in the sun

I played with the knife and managed to cut my foot. Luckily it was only between my big toe and second toe. Screaming loudly, I woke my mother up. She quickly picked me up and ran to a doctor's house who stopped the blood and put a couple of stitches in my right foot. I still have the scar today.

# Shubert's Marche Militaire

I remember being scooped up into my mother's arms from my cot when I was one or two years old. She would carry me around the house, singing along to the old cheap radio in the kitchen that played Schubert's Marche Militaire. I loved her beautiful voice; it was so delightful. This young happy memory has given me a lifelong love of classical music. I have enjoyed listening to the ABC [Australian Broadcasting Commission] throughout my life. ABC is now celebrating 90 years of radio broadcasting, having started when I was a baby, in 1932.

I loved going crabbing with my family and friends in South Perth. Needing only a good set of clothes, washing baskets and a kerosine lamp we headed down to the Swan River, usually just before sunset. One would carry the lamp and the other would pull the basket behind, walking through the shallow water, scooping up the crabs with the baskets. I never participated in catching the crabs but enjoyed walking in the water following them along. The folks on the beach would cook the crabs in large kerosine tins, filled with water, over a small fire built on the sand. The crabs were placed into the boiling water to cook.

The delicious meal of freshly cooked crab on a bread-and-butter sandwich was wonderful.

## *Sydney 1935*

When I was about four years old, my father was sent to Melbourne to learn about the advertising trade before eventually being transferred to Sydney to set up the advertising arm of the West Australian Newspaper (in the Kembla building in George Street, opposite Shell House, which features later in my story). The manager of the Victorian office was Austin Robertson Senior (1907-1988), a well-known football player in the 1940s.

My father was responsible for setting up two new offices, one at St Kilda in Victoria and the other in Sydney. Dad had been appointed the Sydney office but we stayed in Melbourne while he organised staff and offices. My family relocated to Sydney in 1935. We travelled via St Kilda, taking us three weeks to get there. He managed both St Kilda and Sydney offices, handling all the east coast of Australia's advertising for the Perth Daily News. He remained in this role until the 1950s. For his appointment, our family was given a house to rent in Mosman Bay.

At 10am on the 19th March, 1932, the Sydney Harbour Bridge was opened for the good people of NSW by Premier Jack Long. We crossed it for the first time in 1935. The Sydney Harbour Bridge has since become my lifelong obsession, researching anything relating to the bridge build. I now have a huge collection of bridge memorabilia in my home.

### *Mosman Bay*

We moved in and out of three houses at Mosman Bay. Our first home was a newly furnished house at 1 Rose Crescent, Mosman Bay where we lived until 1936. Our next move was into a two-storey house at 10 Mosman Street. It was at this address my family became firm friends with Ada and Leonard Lawrence, who lived upstairs. Ada was Canadian and became my mother's closest friend. Their daughter Anne Lawrence was two years older than me (at age eight). I idolised her and she became my first romance.

### *Luna Park*

The recently built amusement park, Luna Park, at Milsons Point Sydney, opened on the 4th October, 1935. Dad and I were guests of Sir Warwick Oswald Fairfax (1901-1987), who was a newspaper proprietor, to the opening day. We enjoyed free tickets to all the rides. Around this time, I had a Hornby train and two model cars that I enjoyed playing with. There was a stuffed monkey too, of which I have a photograph of me holding the monkey.

My lifelong friend, David Johnson, lived next door at Flat 1, 3 Rose Crescent. As youngsters, we spent loads of good times exploring the wild bushlands of Mosman Bay. I recently visited him in Sydney when his wife died. David made a fortune by bottling Sydney city water after he had filtered it three times.

# 1ˢᵗ Mosman Scout Group

David and I joined the 1st Mosman Scout Group, Wolf Cubs. Ian Pilz, a good friend, was already a cub. I wanted to take charge of our little group of cubs that we formed in groups of about five. We were designated a colour. We were in the brown group but David was in the red group. By 1939 World War Two was looming and Len joined the Airforce while Ada, moved to Balmoral. We moved again, into 4 Park Avenue, Mosman Bay. During this period my father became quite prolific at painting; he also began trying his hand at pottery.

Wolf Cub at the 1st Mosman Scout Group (1937).

1st Mosman Scouts Barn at Mosman Bay.

Some 70 years later, whilst living in Tasmania, my friend and fellow cub, Dick (Richard Geeves) and I were reminiscing about our scouting days when he promptly left the room and returned with his complete Scout uniform to show me. I travelled with Dick from Tasmania back to the Barn in Mosman to celebrate the 100th anniversary of our 1st Mosman Scout Group. It was wonderful to reconnect with my old buddies.

## *Learning about music*

My mother took me to a concert by the pianist and composer Isador Goodman (1909-1982). He played La Campanella and the Ritual Fire Dance. After the concert, some of the audience were invited to join him over a cup of tea, which I don't remember, but I do remember the gaunt-faced tall man playing magnificently. During the early war days when we still lived at 4 Park Avenue, I loved listening to Mrs Koenig, a former concert pianist in Europe. Mrs Koenig shared our house, living upstairs and teaching music. I'm grateful for my early exposure to music, as it has shaped my enjoyment of this genre.

I loved going to the picture theatre known as 'The Kings' on Spit Road in Mosman. I could walk there from Mosman Bay. The Kings Picture Theatre first opened in 1937, a beautiful art deco theatre designed by architects from Crick and Furse. Most Saturdays I was given six pence and one penny for the tram ride but I usually walked and bought an ice cream for one penny.

*Dentist treadle drill*

As children, we were given regular dental check-ups in front of the whole class. The dentist would use a treadle drill that was powered by a foot pedal. The drills would remove tooth decay and fill the dental holes with amalgams. I was intrigued, at a young age, by the workings of this machine. Those amalgams I received as a child lasted most of my life.

*Dad's Sydney Friends*

Ian Idriess (1889-1979) soon became friends with my Dad as Ian's mother lived in the same block of units as my family. Living nearby, he would visit his mother in between his travelling adventures. Her hallways were covered with New Guinea artefacts. My dad would enjoy a cooled beer with Ian, sitting on our front steps. Sometimes our family would join Ian and his mother for afternoon tea. Ian became a prolific and influential Australian author. He published 50 non-fiction books in 43 years, detailing his extensive world travelling experiences, particularly during his time in Papua New Guinea. I remember what a nice man he was, so interesting to chat with, and now I feel very lucky to have met him. I have only one of his books now but before I left Tasmania, I had everyone, all signed personally by Ian. I hated leaving all his books behind. I've kept some framed photographs and pictures signed by him that are in my collection now.

My father walked along Bantry Bay Road, Frenchs Forest every day. He noticed a little shop renovating white goods and bought a refrigerator from them for our family. Gerry Harvey helped organise the delivery with my father. Ever since then, they enjoyed a long-standing acquaintance. Later Gerry became business partners with Ian Norman (1939-2014) and opened their first store together in 1961, the beginning of the Harvey Norman empire.

By now I was starting to notice that my parents' marriage was becoming distant; we always went on separate holidays. My mum took me horse riding in the Blue Mountains and my dad would take me to Bundeena for seaside holidays.

The Japanese Ambassador in Sydney somehow befriended my Dad in 1940. I still can't remember his name. He wore a frock coat and a black top hat when he came to visit. He gave me a model of a Samurai helmet that I still have today. I was very impressed by him as he was so polite and such a gentleman. It was the first time I met a foreigner. He came to visit my Mum on occasion and would take us for drives around Mosman. One day we went to Bradley's Head and he took a bit of footage for home movies when we frolicked at the picnic site. There he promised he would show us the movies. He had taken many videos of Sydney Harbour with his camera. After the war was over the locals concluded he was a phoney, naming him as a spy. He was committed to a period in a Japanese Gaol.

## The Montsalvat Artist Colony

Whenever my father wasn't working, he was painting. He was self-taught and soon got to know the people who ran the Montsalvat Artist Colony at Eltham in Victoria, during the 1930s. One day, somewhere around 1940-1941, he took me to Montsalvat. The colony was on five hectares, established in 1934 by Justus Jorgensen (1893-1975). It grew from an empty paddock to a huge castle-like building with various outbuildings. If you wanted to be there you had to contribute to the community by helping construct the buildings or gardens. There was a constant flow of artists. The dining room was occupied each night by an artist defining art and its benefits. I remember sitting at the head of the table in the huge dining room hall intently listening and learning. One could learn wood-carving or watercolours, for example.

I had not painted or had much of an interest in art but often watched my father paint at his studios. Justus took a bit of an interest in me and introduced me to Mervyn Skipper (1886-1958), a writer who was kind enough to give me a copy of his book 'The White Man's Garden' published in 1930. He introduced me to his wife, Sonia (1918-2008) whom my father had befriended. To this day I think that he was a bit sweet on this young woman. He did a beautiful painting of her as she was doing some sculpture work on the building. My sister, Sally Lynch, has this painting in her collection. My father made two trips down to Montsalvat and my mother was a bit suspicious of what he used to go there for. We later found out, yes, he was two-timing her.

My mother was not happy about me going to Montsalvat, because of the sex aspects. Jorgensen had many female associates and bed mates. He was Norwegian and highly attractive to women. He treated me like an adult. We discussed topics of art and artists, exploring my likes and dislikes. I learnt to appreciate Paul Gauguin (1848-1903), Vincent van Gogh (1853-1890) and Pablo Picasso (1881-1973) in the later stages of my being there.

I grew up with Matcham Skipper (1921-2011) who was a self-taught metal jewellery designer. He made a special piece of jewellery for Princess Diana during the 1980s, as a gift from Australia. I was introduced to him by Mervyn, his father, at Montsalvat. He was a fraction older than me. I didn't meet him again until 1946 when I went down to Melbourne for a

Scout Jamboree. Matcham lived in a garret (attic) of a house in a Melbourne suburb with his wife and son. I often sat at his workbench where he created his silver jewellery. During the 1960s I started making jewellery also, remembering what I'd learnt about silversmithing from watching Matcham as a young man. In 2011, I had a lovely week in Melbourne going to Eltham again, where I met up with Matcham. We spent the day talking and drinking wine together. He invited me to a party, which unfortunately I didn't get to. I regret not attending now, as he died only weeks later.

I have a couple of pen sketches my father drew. Sadly, my father destroyed many of his paintings as he was never satisfied with his creations, so I don't have many in my collection today.

Sir William Dobell frequently visited Montsalvat and was later appointed a war artist for the Department of Defence. His works can be seen in the War Memorial in Canberra.

Margaret Olley (1923-2011), frequented Montsalvat, another well-known Australian painter. The only painting my father ever exhibited was included in a Margaret Olley exhibition. It was held at the Mosman Town Hall during the 1940s. Dad's painting was called 'A Dead Fish of a Place'. Sally has this painting now in her collection. She is a very good artist herself. She has one sketch Dad painted of her doing a woodcarving. Sally lives in our parents' old house in Sydney.

I painted only briefly following my visits to Montsalvat, however my father continued to paint and did many oil paintings. I was taught by my father mostly, and he gave me his easel and supplies. I loved painting trees and wildlife.

Later I became friends with Justus's son, Sigmund Jorgensen (1940-2019) who was born at Montsalvat. Sigmund published a 300-page book, 'Montsalvat, An Intimate Story of an Australian Artists' Colony' via Allen & Unwin in 2014. There is a photograph of the old stables in the book showing where my father and I slept when we visited Montsalvat.

Many years later I wanted to use the great hall at Montsalvat to exhibit tools of interest, however, it didn't work out as Jorgensen would not approve it. We kept in touch and I received many nice letters from Jorgensen over the years. My association with Montsalvat came and went over the years. Jorgensen had some sons who continued to run the art school. Montsalvat is no longer an artist colony, having been heritage listed on the Victorian Heritage Register. It is now a tourist attraction preserving its historical significance for future generations to enjoy.

# World War II

*World War II lasted six years and one day, commencing on the 1st September 1939 and ending on the 2nd September 1945.*

Aged about ten years, I became a Warden alongside my father, the Chief Warden, during WWII. When the sirens went off, it was our responsibility to walk up and down the streets, telling people to turn off their lights and pull their blackout blinds down, in fear of bombs being dropped by war aeroplanes, before heading into the air raid shelters. Now and then we would have to practice by rehearsing our protocols. People would often yell at us for following our designated Warden duties. My father wore a metal helmet and I felt quite disappointed at the time as I didn't get to wear one. Thankfully there weren't any bombs dropped in Sydney.

At the Perth Daily News, the manager, and my father (being the eastern states manager) were called up to join the war. However, only the general manager was accepted as my father had flat feet. He was then promoted to the general manager's position. This required him to move back to Western Australia to take up that post in Perth.

## *Voyage to Fremantle*

Firstly, only Dad and I moved to Western Australia. I was about eleven years old by then. We sailed on the *SS Gorgon* (built in 1908) on a 15-day voyage. This ship was privately owned but was commandeered for the war efforts. It sailed non-stop from Sydney to Fremantle, cruising around the southern coastline of Tasmania. We sailed past Cloudy Bay on Bruny Island, a place I would later call home, many decades later. I didn't enjoy the open seas, as I became very seasick on this voyage. The ship's doctor wanted to operate on me for appendicitis, to remove my appendix. Thankfully Dad locked the cabin door so he could not get in and I eventually recovered from my sea sickness.

Just after we left Sydney the ship was fitted with paravanes (mine sweepers). They were hung out on each side of the ship, torpedo shapes gliding through the water to collect floating mines for detonation. Any nearby mines would get caught on the cable of the paravanes and when the cable was cut, the mines would float to the surface where they could be destroyed by gunfire.

During our voyage, the Japanese joined WWII and caused havoc when their submarines entered Sydney Harbour on the 31st May 1942. Mum was packing up our home at this time, in preparation for our move to Perth, when she experienced much shelling in Sydney Harbour. A torpedo landed in a park near our home, and the Mosman jetty was damaged during an attack on Sydney Harbour. We were carrying ammunition to Western Australia and were completely unaware of this attack until we arrived in Fremantle.

Dad had a cousin on board, Elizabeth Durack (1915-2000), who later gave me a copy of her book, 'The Way of the Whirlwind', published in 1941. Both Elizabeth and her sister, Mary Durack (1913-1994) became well known for their literary contributions, writing about the NeverNever lands (Australian outback) and the Mirriuwong-Gajerrong peoples. I didn't get the opportunity to verify whether they were my dad's cousins or not, but I had all their published books.

# Western Australia

Upon arrival in Fremantle, we were met by Mary Durack. Dad and I stayed at his sister, Clarice 'Clal' Henderson's, house in Cottesloe. Later we all moved to a rented house in Queen Street, South Perth, belonging to a man called A.G. Johnson. Our next house was at 5 Darley Street near the Coote Street ferry and my school, Wesley College.

### *Swan River Crabbing*

Crabs were abundant in the Swan River so it was incredibly easy to catch them. After boiling them up in our kerosine tins on the sandy banks, when the people came off the Mend Street Ferry from town, we would sell them our crabs. The crabs were sold according to their size, which we gauged by the width of the timber boards of the jetty. One board was about six inches (15cm) wide. We charged three pence for one board width and four pence for a 12-inch (two boards' width). We made enough money for us all to go to the pictures.

South Perth had only a few houses during the early 1900s as it was a difficult place to travel to but the flour mill is still there today and I remember the hole in the roof of the garage that you can see in many pictures. I made that hole when I fell from the window onto the roof while playing there in 1943. It was not part of the National Trust back then, so it was a free space for us to play around in.

I soon made friends in Darley Street, with many of the neighbouring children. It wasn't too long before I got a crew of them together to form a gang. I was the leader and conjured up the name 'Px Gang'. I was intrigued by the note on a doctor's prescription where they signed and marked Rx. I liked the symbol but decided as we were not medical to change it to Px. We often did things together, like going to the outdoor picture theatre in Como, and of course crab fishing.

I delighted in driving a tram that travelled to Mends Street Ferry. On occasion, I was allowed to drive this tram for short distances whenever the driver extended an invitation.

Tram at South Perth (1943).

### *Dad's artist friends in South Perth*

During our stay in Perth, 1942-1946, a close neighbour, who became Dad's best friend, was artist Lionel Jago (1882-1953). He would visit most Sunday mornings to have a beer and talk. I took his photo in the backyard of our house. He gave me one of his paintings, which I still have today. I recall his wife, Christina McTavish, quite noticeably as she always used light makeup, resulting in a very white face to prominently feature her bright red lipstick. She showed much interest in the history of paintings. After Jago died, I bought a self-portrait from his estate that was being auctioned.

Lionel Jago in our backyard, I took this photo.

Leith Angelo was an illustrator from the Perth Daily News. He lived in Perth during the early 1930s. He gave my father two pencil sketches of South Perth, one being a cottage and one of the Old Mill. I inherited these but they were sadly stolen from our house in Jersey Road, Woollahra, NSW, around 1965.

# Erecting Camouflage Tree, Menangle (1944)

Sir William Dobell (1899-1970) was a portrait and landscape artist. He won the Archibald Prize on three occasions and was knighted in 1966. In 1942 he was appointed an Australian War Artist and his commission was to capture the war effort in Western Australia and in particular war camps where aircraft were stored.

At work, my father met William. He was such a likeable person that upon finishing his wartime commission, my father invited him to our art studio at home, to view my Dad's paintings and have a meal with the family. They got on very well and my mother liked him too.

He saw one of my paintings that Dad had framed and hung on our wall. Bill asked, *"Who is the artist?"* and Dad replied, *"My son, Terry."* Overhearing the conversation, I asked Bill if he liked it, and he said, *"Yes."* Proudly, I suggested he have it with my compliments. He seemed happy to accept it and said he would send me one of his paintings in return. The painting I gave him was of a koala sitting on the fork of a gum tree. I would love to find this painting again but have no idea of its whereabouts.

Two or three years later, the friendship between my father and Dobell grew to where they had constant contact with each other. One day they were in Bill's studio in Victoria, and he said to Dad, *"By the way, I owe Terry a painting."* Bill explained, *"I had to redo this picture because there's a figure in it who's a little bit too big and too fat, and people didn't particularly like it, so I'll send it down. The copy I've made is now in the War Memorial Museum in Canberra."* His peers judged it as being a bit rude with the foreman's stomach hanging out over his shorts, so he had to repaint this scene with a slender foreman, therefore, he didn't put a lot of value in the painting he gave me. The painting shows a team of men pulling on a rope to erect a camouflage net around an aircraft at an airbase. The painting was completed in Western Australia and is called 'Erecting Camouflage Tree, Menangle (1944)'.

I did visit Bill many years later, at Dora Creek, when I was then living in Taree, NSW. Dad had always remained close friends with Bill having their connection to art.

Many years later, when I was a young man travelling the world, my mother cared for this painting until my return from England and Canada in 1960. Back home in Australia I hung it on the wall and my newly-wed, Jacquie, said, *"Take it down, I can't stand it."* It was then put on top of the wardrobe where our cat used to sleep on it.

A few years later, when the opportunity came to buy a house in Vaucluse, Jacquie and I needed the money for a deposit. Jacquie thought, if he was such a famous man, maybe his paintings might be worth some money. I took the painting along to a gallery for an opinion. They said, *"You're going to have to have it cleaned,"* so I took it to Hal Missingham (1906-1994) who was the Director of the NSW Art Gallery. He was an old chum of my father and I was most grateful because Hal happily cleaned up the painting, free of charge. When I took this painting back to the art gallery they said, *"Ah, that's better, we can sell it now."* They placed two spotlights upon it for focus and it was sold during the early 1960s to a well-known Brisbanite, a Major, who was a collector of Dobell's paintings. He paid the considerable sum of $10,000 which covered the deposit on our Vaucluse house. Bill's only remark later, when I told him how much I had sold it for, was that I should have got a lot more, but he didn't blame me for selling it to buy a house. We purchased the house for $25,000 (circa.1960) and sold it for $1.2million (circa.1970).

## *Roget's Thesaurus*

Stanley Smith (1907-1968), a journalist and humanitarian, was among the many visitors to my father's office at Perth Daily News. Stanley made an appointment to see him to discuss how the newspaper could assist him in supplying chopsticks, bowls and blankets, etc., to people in the Pacific islands who had been badly affected by the Japanese invasions. They seemed hit it off and after a few beers at the local pub, Dad invited him to visit our house. I only have vague memories of this first visit but I do know he was a very pleasant man and my mum thought he was very nice too. One day he revisited our house, taking me aside, as we talked a bit about the war and its effect on my education. He then presented me with a book. I strongly remember his words: *"Although my learning was limited, I should always, when writing, never use the same word twice."* The book was a Roget's Thesaurus, to help me look up words to suit a situation without repeating myself. After that Stanley was gone, thinking I would never hear from him again. This book has always been very close to me in the house, as I like referring to it whenever I have some writing to do.

Moving forward, in my travels some ten years later in 1955, I lived in London with my pal, Ian Hamilton. Together we were meeting my cousin, Bill Angove, for a pub lunch in Soho. As we chatted on the pavement, I noticed a big blue Rolls Royce stop beside us, only for a few seconds but long enough for Bill to recognise the passenger in the back seat: Stanley Smith. The car moved off down the road. I recounted to Bill my meeting with him in South Perth back in 1945. Bill reminisced he had a photograph of him and Stanley's daughter when he was working for Cecil Beaton. Back in 1954, Bill explained that Stanley had thrown a party in London for his daughter's 21st birthday, which was apparently the most expensive party for a private citizen in London since the war. Thinking that was the end of that, I promptly forgot all about my connection with Stanley until 2023 when I heard about the Stanley Smith Trust. What was that all about, I wondered? With the help of my son-in-law, Benno, we tracked down a book of his life written after he died. The trust lives on and his book 'No Substitute for Kindness; The Story of May and Stanley Smith' published in 2017 arrived; here was the whole story of that remarkable man who had achieved so much in his life. I was so privileged to have met him. Our meeting of course is not mentioned in his book, as there was no one to record his trip to South Perth in 1945.

Alan Trevor in my backyard, South Perth (1944).
He became a famous stage actor and I was featured in a radio serial on ABC with him.

## The War Ends; Back to Sydney (1946)

Dad was recognised for his war effort at home in maintaining the free press. He was honoured by the Duke of Gloucester, who was the Governor General during WWII.

The war was over and the general manager came home to reclaim his position. Dad was demoted to the eastern states manager's role and was expected to move back to Sydney, promptly. It was apparent during this time his marriage to Mum was over. She decided to go back to Sydney too, to get away from Dad's relatives in Western Australia no doubt. She was looking forward to seeing her old Sydney friends again. I left Wesley College and packed up all our things.

About a month after Dad went to Sydney, my mother and I planned to travel aboard a train to the east coast of Australia with all our furniture. The journey started ok but upon arriving in South Australia, heavy rain flooded Lake Ayre, which caused sections of the railway lines to be completely taken out. We became stranded for two or three weeks, with all the other passengers, in the small country town of Tarcoola. There I spent much of my time working in the local bakery. The oven was sometimes called upon to cook a sheep due to the difficult times. Conditions were pretty rough as we had to sleep on the floor of the train or on the upright seats.

Once the railways were mended, the train left late at night so we could not see the flimsy trestle bridge I had witnessed earlier. We eventually arrived in Melbourne for a stopover and saw Dad in Collins Street. My mother wouldn't talk to him. Sometime later, in Sydney, Ada Lawrence had invited my father to a party she was holding. It was there that he met the vivacious Frances Murdoch.

Getting into Sydney, Mum and I had to search for accommodation, which was very difficult to find after the war. But we managed to acquire what could only be described as a 'garden shed' at 9, Clitheroe flats, Reid Park, Mosman Bay. Grateful for the shed, Mum quickly transformed it into a habitable residence.

My parents divorced and after some time my mother remarried to Peter Paterson. Dad had a house in Newport Beach and about a year later, I went to live in his next house at 5 Bradley Heads Road, Mosman. I eventually moved back to my Mum's shed after a year or so of living with Dad. It was an unsettling time for me as a teenager, not knowing who to live with.

Eventually, my mother moved to Blackheath in the Blue Mountains. Her final move was to a village called Faulconbridge, also in the Blue Mountains, where she died at the good age of 102 years.

### *Last year of school*

I went to Mosman High School for my final year. My favourite subjects at school were woodwork and physics, understanding the planes to cut wood, and the mechanics behind the Bunsen burners. The teacher said to me one day, "*You can be the tool boy and make sure the tools are sharp, clean and ready for use because I can see you love doing that sort of thing.*" I replied, "Oh yes I love it!"

Our school was divided by gender and we could only mix during dance classes. I was still very shy, and terrified to touch a girl, but over time dancing did become pleasurable as my confidence grew.

During that time, I rejoined the 1st Mosman Scouts Group where I met the young John Laws, who later became an esteemed Radio Announcer. I soon became a Patrol Leader of the Curlews and John was in my patrol group, we did have a lot of fun. I had set myself on becoming a Scout Master but got to be a Troup Leader before becoming a Rover.

I learnt to play tennis at the Valley Tennis Courts, between Mosman and Cremorne. My two coaches, Lew Hoad and Ken Rosewall became world-renowned tennis players. I belonged to the school tennis team and travelled to other schools on Saturdays to play. After injuring my knee in an accident, I became the umpire for the school's team.

As a teenager I had no money; my parents were divorced and mum was struggling. She was working at the Abbott Laboratories in Botany as a supervisor in the production line of manual workers. I felt I should be paying my way in my last year of school.

One of my good friends, Geoff Harris, went to the same Sydney kindergarten, primary and high school together. His father ran the local post office at Mosman Bay Wharf. Sadly, Geoff was killed by Indigenous Papua New Guineans whilst in Papua

New Guinea during the 1940s while working as a Patrol Officer. His death was recorded as a mistaken identity.

There were three businesses on the wharf: a post office, a general store with a newsagency and a dry cleaner. My stepfather, Peter, bought the Laurence Dry Cleaning shop and added a florist into the business. Without a car license, I drove to the markets every second day, very early in the morning before school, to select flowers for the shop. The schoolgirls would sell the flowers on the Sydney Harbour ferries.

## Prouds the Jewellers (1948)

During my last year at school, I worked part-time at Prouds the Jewellers, on the corner of Pitt and Elizabeth Streets, Sydney. After leaving school I worked full-time over the busy Christmas period. I took the job so I could pay for the tools, overalls and work boots I needed to start my apprenticeship the following year. I think Mum was tickled pink that I was doing that sort of thing.

I worked in the pen and pencil department. My boss said there was a new line of pens recently developed by Biro that had an ink tube with no filling required. The biros came in many different colours. There was a heavy amount of advertising for these pens, so we had quite a rush of people and sold large quantities of the new biro pens.

One day a very tall man came into the store and asked to see the cigarette cases. It was prominent Australian actor Chips Rafferty (1909-1971). He bought five gold cases engraved with, *"Congratulations on our latest movie, The Overlanders, Chips Rafferty."* This film was released in Australia on the 27th September 1946. The gold cigarette cases were presented as gifts to the cast/staff for the success of the film a year later. I feel honoured to have met this famous actor at that time. The engraving was made with only a small chisel by hand. The store had its own designated engraver and I was privileged to watch him work. In later years I bought myself a set of engraving tools and taught myself to engrave on silver.

The store also sold lighters, I had never seen a lighter up close before working there and it was good to take the lighter out of the box, look at how it worked, and then be able to sell them.

### *The Blue Lagoon Audition*

Jean Simmons (1929-2010) came to Australia to star in the movie 'The Blue Lagoon' in 1949. The producer, Frank Launder, wanted several children to audition. The parts were for two young children who survived a plane crash on a tropical island. They grew up together and had children, and were finally found as adults. I applied for a part for the young boy in the first half of the movie. In my collection, I have kept the black and white photograph of all the boys who applied for the screen test on the day. The director said that I would be perfect for the part but had brown eyes, whereas he wanted a blue-eyed blonde boy for the character. Frank Launder did use me in several other films that were shown in theatrettes, mainly short advertisement films. I had the opportunity to meet Jean Simmons, the lead actress, she was beautiful in real life.

## *Pittwater Explosion*

I was sitting on our verandah, as I was visiting my father at Newport Beach, having a quiet beer when we heard a loud explosion. I looked across the harbour to see smoke over at Pittwater. I didn't think any more about it at the time until the news of the day reported many people had been hurt in the incident. Dr Gilbert Phillips (1904-1952), a brain surgeon, was miraculously nearby being the first on board to safely rescue the family after the explosion. He was a skilled surgeon, having worked in New Guinea when the Japanese were there operating on the prisoners during the war years. My step-father, Peter, was invited to a special dinner for Johnny Walker in Crows Nest and he kindly invited me to join him. This same surgeon, Gilbert, was the president of the Sydney Food and Wine Society. During the evening Gilbert and I fell into a deep conversation about bushcraft, living off the land and cycling. So soon after I found myself accompanying him on bike rides through the Sydney National Parks and the Blue Mountains.

He lived in a household full of women, having three daughters, and a wife, mother, mother-in-law and step sister. All he wanted was to have a son that he could talk to, so I fitted the bill. He invited me to spend a weekend on his yacht with his daughters, hoping I might like one of them to marry and live in his house of women. I didn't manage to connect with any of his daughters although they were all very nice women. In 1952 he developed cancer in his leg and it finally got him, dying that same year. I found him to be a hell of a nice man.

Many years later my friend, Ian Hamilton, married a beautiful girl in London, Beatrice, who was from the family on that yacht. This detail was not apparent to me until decades later, when Ian and his wife, Beatrice, were visiting, she recalled to me how, whilst on the family yacht, moored at Pittwater, they had petrol leaking when her father started the engine and all hell broke out. Her father was badly burnt but lucky for him, a doctor who was drinking at the Newport Hotel rushed down and was able to treat him along with Beatrice, her mother and sister. The doctor was Gilbert and they soon became good friends following this incident. Gilbert later saved her life during childbirth when complications developed.

Beatrice lived in Double Bay and had come to my wedding in London. Jacquie liked her a lot and later she became godmother to my daughter, Sasha. I met Beatrice again in

2000, at my daughter, Belinda's, house, I was struck by her loveliness. We used to keep in touch by writing to each other. After my second wife, Marian, died, we eventually rekindled our connection but after her second look at me, I was quickly put into the friend zone, hands off. We did live together for some two months before I sold my Tasmanian property for my relocation to Queensland.

## No longer an only child

My dad remarried a lady named Frances and together they had their first daughter, Jane. Being around at that time I remember her birth, and grew up knowing her as a toddler quite well.

Jane has become an authority on one of my aunts, our great aunt, Muriel Butcher. We have been discovering a lot of information about her, including digging up old photographs. One in particular I found after her son, Alan, died. It was a moth-eaten picture but Jane went off to a photograph repair specialist, who had it restored fully and the result was marvellous. By sheer coincidence, Jane's husband, David, worked as a photographer for Ian Hamilton at the same advertising agency. Ian retired and David is now working near Cockatoo Island in Balmain where they are restoring the old Sydney Ferries. The photographer taking pictures of the restoration work was my next-door neighbour in Pitt Street, Redfern. It's amazing the coincidences of meeting people.

My second half-sister to arrive was Elizabeth. I didn't get to know her until she was a beautiful young woman ready to marry. She married a factory cleaner who had very ambitious ideals. They bought a house at Cottage Point on the edge of the Sydney National Park near Palm Beach. The house had the most glorious views you could ever imagine, overlooking Broken Bay. They lived there for many years. My Dad would often visit them. He enjoyed sitting on the verandah admiring the wonderful water. After he died, we chose to place his ashes under a beautiful rose bush in this garden directly in front of their verandah, a place we knew he loved. To my horror, shortly after it seemed, Elizabeth and her husband sold their house and moved to the other side of the Blue Mountains. They purchased land from an old company that had developed plans and equipment to dig for oil, but the site had since been abandoned, going to waste. Elizabeth and her husband hoped to turn it into a holiday place exhibiting the old equipment. Their plan didn't work out. It wasn't long before they became separated, having to sell up and move back to Sydney.

Elizabeth soon had another adventure lined up and moved to Alice Springs and became a country nurse. She lived there for many years. Elizabeth was a wonderful woman with a vibrant spirit but we lost contact for some time. Upon my retirement, when we were living in Tasmania, she turned up on our doorstep. I recognised her immediately and she spent the entire day talking with us catching up on everything. At the end of the day, she picked up her bag and disappeared into the sunset. I have never seen or heard from her since. Elizabeth had two children, and I have met their families. As far as I know, she is living with one of her daughters in Melbourne.

I do have a strong bond with Sally, my youngest half-sister, having known her for most of her life. It has been wonderful watching her grow up. Her life has been full of all sorts of interesting adventures. She is happily married to Bryan, an artist himself. Shortly after my step-mother died, and my father had already passed, the family home was bequeathed to the four of us, my three half-sisters and I. Sally decided to manage the money by raising enough funds to pay off everyone's mortgage and divide the balance of the funds equally. Sally continues to live at Frenchs Forest in the family home, which Dad and Frances loved dearly. Dad's pottery shed was still in the backyard where he produced remarkable pottery including some commercial pottery. He created his paintings there too. Sally has become a very successful painter in her own right, having had quite a lot of exhibitions around Sydney. Bryan is talented too, and a very busy landscape gardener. He completed my stonework pathway that leads from my front door to the driveway in an arc across my front garden here at Flaxton in Queensland. I enjoy admiring it every day. Bryan has since retired and has taken up sculpture using clay. He makes huge table-sized models of Sally. I've only seen photographs of these so far. I am dying to see the finished product in person.

Sally has always been a big part of my life, having introduced to me many of her interesting friends. For example, 'Beachie', a world champion surfer back in the early 2000s. I had been reading about her in the newspaper and discussed it with Sally. She told me it was one of her best friends. I wrote to Beachie, and she was enthused that I was interested in her achievements in surfing, following her career right up until she retired. Beachie put out her own movie and wrote a book about her adventurous life. She had a little shop selling surf gear at Lane Beach. We got on well together, a lovely person.

Another friend of Sally's was Claudia. When we were living in Tasmania, ABC created a program called 'The Collectors'. Being an avid collector, I became quite a fan of this program, having written to a number of the members until I saw them on a knowing basis. One of the reasons was the location of the program; it was produced nearby in Hobart. Over the years I got to meet all the people on the show and was promised they would come out to Cloudy Bay one day to have a look at my extensive tool collection.

Sadly it never happened because the young fellow who was the master of ceremonies (MC) for the program ran into a lot of criminal trouble and was swiftly dismissed by the ABC, later being jailed for his crimes. The show paused for a time then re-emerged with a new MC, a young woman who didn't last long either. Then along came a lovely young woman, Claudia Chan Shaw, an avid collector herself. She hosted the show until the program ceased production in 2011. She became very well-known and currently hosts a new program called 'Antiques Downunder' which commenced production in 2022. When I travel back to Sydney with my daughter Sasha, we always go out to see Claudia and her husband at their house. We enjoy having a few drinks and looking at their collectible items. I'm pleased Sally and Claudia have remained good friends. Some years ago, Sally painted Claudia and submitted it to the Archibald Prize. It was recommended but didn't win the major prize. Being a little biased, I did feel it was a wrong call. It was the best painting there, but who am I to judge such a competition? I have one of her pictures here at home.

Every year Sally hand paints a series of postcards. I always love receiving a gift card from her. My family are scattered around the country and unlike many other families, we were never able to all get together over Christmas. I'm still missing the other two girls, Jane and Elizabeth.

# CHAPTER 5

# Engineering Solid Foundations

### Fitter & Turner Apprenticeship (1948-1952)

In early January 1948, I started my apprenticeship with Coote and Jorgensen in Botany Street, Alexandria as a fitter and turner, signing up for five years. Mr Jorgensen, the engineer, spent a lot of time on the factory floor with the workers. He was a friend of my step-father and was a nice man to work for. My first year's wage was 15 shillings a week and my lunch cost tuppence, which was quite expensive so I made sandwiches. I bought a pair of bib and brace blue overalls for my uniform, which had to go to a special cleaning company because of the grease and caustic chemicals on them. I bought tools from a travelling van. The company provided some tools but I had to make my cold chisels.

Mr Jorgensen took me to the maintenance room on the factory floor. There were two work benches in the middle of the room. I was asked to build a surface plate in the shape of a rectangle made out of cast iron. I had to scrape the surface until it was perfect, using a perfectly flat plate covered in Prussian blue. The colour was transferred to the high spots on the second plate, to identify the elevated spots to be filed off. It took three months to do. I was told it was a good job. Then I was given a job to make gauges with U-shaped pieces of steel, one side 'go' and the other side 'no go'. Every lathe operator required one for each job he was doing. I eventually required gauges for my work too.

The most important part of my job was tempering the gauges. I used a ball pein hammer (a small hammer with two heads, one round and one flat). We used an air supply for cleaning, pipes everywhere. Many machines ran from a belt in the grinding department and every machine had an air chick. On one occasion someone had pulled off a pipe disconnecting the air supply of the chuck. To fix it I turned the air off without warning. I got into a bit of trouble over that incident, being quickly reprimanded.

There were 150 machines on the factory floor including various lathes, grinding machines and gear-cutting machines for all types of gears (lipoid, gearbox, and other industrial applications). Sam Horden, the foreman, moved me from machine to machine, explaining how each one worked. He supervised me for competency before he left me on my own. We were very polite and friendly, saying good morning and good night daily to each other. The machine to learn was a 6-foot diameter gear used for the sluice gates of Warragamba Dam. Teeth for these gears were being cut.

### *Queen Elizabeth and Prince Phillip's Visit*

While I was serving my apprenticeship, on occasion I got a lift from one of the workers. This fellow had a vintage Packard (a classic American luxury car). It was in very good condition and he always kept it spotlessly clean. One morning, as we drove through the Sydney suburbs, we were stopped by a policeman. He told the driver, *"Pull over to the curb, park and wait,"* as he explained the recently arrived Queen Elizabeth and Prince Phillip would soon be passing by. There were four of us in the car that day. We all got out to stand on the curb. Soon enough, a police car went by followed by a car with the royal couple. As we were the only people there, their car slowed down while we waved and cheered. The royal couple responded and waved back. Little did I know at the time I would be meeting Prince Phillip in person, when I was working in Canada a few years later.

After four glorious years, the day arrived when I was told my next job was going to be in the drawing office. Around this time, Coote and Jorgensen had sold out to Borg Warner, located at Fairfield. I had to move there as it would have been too far to travel to work. Fairfield was mainly bushland back then with just a single factory or two dotted around in the large paddocks. About four hundred people worked at Borg and Warner.

Being in the office for the next phase of my apprenticeship, I had to wear a suit, polished shoes and have clean hands. My toolbox was locked up and taken home. My next set of tools was the drawing instruments that I still have with me today. I loved this work. At school I had enjoyed technical drawing, finding it quite easy, so I was prepared for this job. The drawings had to be one hundred per cent accurate. It was nice to receive the drawings back from the boss with *"perfect job"* written on them. I made good friends with other draftsmen and some of the women. One girl I made a special friendship with; we went dancing and enjoyed picnics together.

My last job as an apprentice was in the final inspection department. That was an important role. If you found a fault, it was either human or mechanical. Most times it was a human fault, from not cleaning appropriately or an upset with the settings. I would go along to explain how the fault was made, telling the worker how to fix it – being firm but friendly. I was never nasty to anybody as it wasn't in my nature. I enjoyed the time I spent with them until they got it right. Sam said he liked the way I handled people over the problems found. I had been through the same issues myself getting very ticked off many times over the five years and was more than happy to show my experience in overcoming such things.

Many years after my apprenticeship, when I was working for a packaging company, we packaged a set of gears mainly for differential. It consisted of a crown wheel and pinion. We used to put the pinion on top of the crown wheel, on a wooden backing secured with wire then dipped it into shellac to seal against moisture.

Borg Warner later manufactured car gears and differentials for Ford and Eaton Trucks.

Years later they were still packaging that way until one day the Japanese came along with a new solution, vacuum packing onto a plastic card. The company started to think about this way of packaging. My boss got a telephone call from Sam, now the managing director of Borg Warner, to go out to introduce the new packaging process. I said, *"I know this bloke, he was my foreman, during my apprenticeship."* I went out there with the sales manager as it was going to be quite a big order.

We discussed the new concept and when the meeting was over, Sam leaned over and said to me *"Whatever happened to you, where did you go?"* I replied, *"I didn't say goodbye to you on my last day as you weren't in. For the record, Mr. Rose, being my boss at the time, was head of the final inspection. He had given me my letter of apprenticeship completion, and I then proceeded to say goodbye to my fellow workers."*

Back then, there was no such thing as a farewell party or anything of that nature and I just went home. I inquired why he wanted to know what happened. Sam disclosed he wanted to promote me to shop manager, needing someone to take his job, as he was being promoted. Sam had already spoken to the management with my recommendation for the role but didn't want to tell me too soon in case I was not selected. I was proud that he had such faith in me. He was pleased with my progress within the company, having learnt to do the physical jobs but also being able to get on well with the workers in the factory as an inspector.

I was chuffed that he remembered me and had many compliments for me. At the time I was disappointed this promotion didn't eventuate, but over the years I came to terms it was not meant to be, as I didn't ever want to make a career out of working in a factory. It was good fun but the world was beckoning me to explore and learn much more.

Driving back, the salesman said to me, *"Oh, you made quite an impression on Sam, they have given us quite a big order."* I didn't think it had anything to do with me; Sam was impressed with what we had to sell and how it was presented between the two of us. We had simply explained that you put the gears onto a piece of cardboard, and push a button to seal it, all made out of plastic. The old way, took a lot of time to package it and when it was unpacked, it was dipped in methylated spirits to wash the shellac off.

### The Oldsmobile Yank Tank

A friend and I bought an Austin 7 for £10 and we were doing it up at his place, to get the car working. His father had a big American car; it could have been an Oldsmobile. He sold it with the idea of buying two smaller cars for himself and his wife. They chose a Volkswagen, at the time. This day he came up, as we were working on our car, and he said, *"I will give you some money to clean the Oldsmobile before I take it to the buyer tomorrow."* We accepted his offer and as we were cleaning the car, a couple of friends came over to see what we were doing. When we had finished by about 5pm, my friend, who wanted to drive the car, said *"Let's test drive the car to Frenchs Forest, to make sure everything is working ok."* We all agreed and jumped in.

We got out to Frenchs Forest and crossed the Baringa Highway when one friend, Trinity, pulled up alongside and said, *"I will race you to the Harbour Bridge,"*. Being of a young age we were in the big Yankee car, nobody was going to advise us not to. There were three of us in the front seat and two in the back seat. I was in the front next to the passenger's door. We were halfway between Frenchs Forest and the Spit Bridge, where there was a sharp right-angle bend. As we approached this corner, the driver sped up and we went around the corner broadside. The car then rolled over a couple of times. My passenger door flung open and I catapulted out. The car continued to roll before coming to a stop at a telegraph pole. The boys came back to check on me, getting me to stand up, although I had no feeling and I was completely buggered. The other car arrived soon after. I was speechless, I couldn't think, I was in a lot of pain, I was in distress, and all that sort of nonsense. It was decided I needed to get to the Mater Misericordiae Hospital at Crows Nest but by the time we came up Spit Hill towards Mosman, I began to moan and groan a bit, so they told me. They changed plans and took me to my family doctor on Military Road. They eventually found his clinic, took me into the waiting room, and hastily fled off. Sooner or later the doctor came out, inquiring what had happened to me. He swiftly organised an ambulance and I remember very little from that point on.

Three weeks later, as I was slowly recovering in hospital, my mother came in to visit me. My face was now one big scab and in Mum's horror, at the sight of it, I believe she was given a tranquilliser and then sent home to calm herself.

Upon discharge, about a week later, it was recommended I go and stay with my mother until I fully recovered. It took another week or two before I started to regain my strength, becoming fully able and physically better. There were no broken bones but my watch was ripped off and there is still a scar where my thumb was hanging off backwards.

I phoned the driver of the car several times, about my progress, thinking he would be interested. The driver informed me nobody else got hurt in the slightest. He said I shouldn't have opened the door, trying to get out. I didn't believe that I did open the door. I have no recollection, as I was hanging onto the fellow who was sitting alongside me as we sped along.

Sometime later I was on the golf course doing nine holes at Blackheath with my friend who had driven the car. I told him I was now doing ok. He said, *"You lucky bastard."* Meanwhile, my step-father had investigated the accident and found it was caused by irresponsible reckless driving. The car was a complete right off. They had no car insurance because my friend's father had cancelled the policy due to the proposed sale the following day. I was completely unaware my stepfather had made a compensation claim. He was successful in this claim, suing the driver for a very large sum of money, but I never saw any of that money. My step-father had kept the entire sum to himself.

# Royal Australian Navy Reserve (1949-1955)

Sometime during 1949, while still serving my apprenticeship at Coote and Jorgenson, all apprentices were summoned to the dining hall for a lecture by the Department of Defence. It was expected that the war in Korea was expanding and although we had troops fighting there, more were needed to protect Australia. There was a need to expand the Australian Defence Reserves. We were asked to apply and were taken to a hospital for a medical checkup. I applied for the Air Force Reserves as my number one choice, hoping I would get the opportunity to learn to fly. I got through the medical but failed the eye test, however I was promised a good role in their ground crew. No thanks, I thought to myself, so I went back and applied to join the Navy. I promptly received my uniform, cap and shoes, and was told to report to the training school at the Navy Depot in Rushcutters Bay, Sydney.

In the Reserves, I reported weekly, every Friday at 6pm and Saturday at 9am when required. My first course was learning to march up and down the parade ground. Several weeks later only one hour was devoted to marching, the rest to being taught seamanship. This involved mainly learning to tie knots with a rope, the same as I had learnt in the Scouts. I wondered where could you use knots and rope on a ship. However, we did learn the layout of ships and a new language for living on a ship, such as: 'deck' for floor, 'cabin' for bedroom, 'the heads' for toilet and 'dobi' for laundry. Other names like 'scuppers', 'hatches' and 'port holes' all became part of our conversations. Any Officer had to be saluted on the parade ground. At all times we wore our caps and kept our black shoes highly polished and shiny. I liked the discipline of the Navy because it all had sensible meaning, and when the order came to clean your cabin, they meant it.

During my time in the Navy Reserves, every year I had to take a week off work to do sea training. One year I went to Grafton for a Naval anniversary with three of our local boats for the celebrations. Another year, we headed down to HMAS Albatross. We were having sea trials about 40 to 50 km off the coast of Sydney. One day the alarm bell rang for all hands to stations in the engine room. We began pursuing a Russian or Korean submarine and were going to be running at full speed. We were in international waters and could see New Zealand in the distance.

My job was to get underneath the pistons. The piston rod needed manual lubrication and I had to get up real close to it with a piece of iron and a rag on the end, dabbing the piston with coolant on every stroke to keep it cool. One particular piston was giving us trouble. There were three sailors on my side and we had all been working on it for quite some hours, doing the same job, when the propellor shaft on the other side seized, stopping the engine entirely. The main bearing had melted. To gain access, there was a square cube from the deck going all the way down to the hull of the ship where the bearings sat. The Engineering Chief and I went down there and found it. We removed the two halves to recast, polished the white metal and then cast a new bearing around it to lock it up. It was only halfway through our repair job when a depth charge (anti-submarine warfare weapon) was released and exploded in the water. The Chief immediately called the Bridge, furiously taking the Officers to pieces about this incident.

By now we were well into 8 hours of work but I was feeling quite proud to have achieved such a complicated job. We did get the ship running again but figured the submarine was long away. Now it was all hands on deck as we were near New Zealand and out of our continental waters. The Officers exclaimed that if the public knew about this incident, people would panic, demanding the crew to swear not to discuss it with anybody. I adhered to this confidentiality for a considerable amount of time before talking to anyone about it. It was hard to keep such an adrenaline-fueled mishap a secret.

Over six years, from 1949 to 1955, I went from a Seaman in an engineering branch to an ERA (Engine Room Artificer). I changed my uniform to black, had more exams, and passed ERA4. Then one day I was called before the Captain, and he said *"You have done extremely well in your time here. We are going to ask you to join an Officer training class, and you will be required to wear a square rig again."* I learnt how to use a pistol and rifle. My marching and arms rifle all had to be perfect.

### *Naval Trade Test*

Coote and Jorgenson gave me a couple of days off to do a trade test at Garden Island. I arrived there from Fairfield at 7.30am and presented myself at the office. The test was as follows:

- I was given a length of hexagon brass 3 inches (7.62cm) across, and flats, 15 1/2 inches (39.37cm) long.

- With this, I had to smooth off on the lathe a piece of 3 and 2/5 inches (8.36cm) from one end.
- From the remainder, I had to cut a washer 1/4 inch in diameter with an outside diameter of 2 and 15/16 inches (7.46cm) for 10 inches (25.4cm).
- Then turn the entire cut into a 2 inches Whitworth thread. Clean and dress the other end, leaving a hexagon section of 2 inches.
- From the short length dress both ends and bore a hole through the centre and then turn a thread 2 inches diameter inside, thus creating a hex nut.
- The two threads had to engage after filling the nut to the bolt section was placed on the bench vertically or standing up on the end.
- The nut was to be given a sharp twist and on its own gravity slide down the full length of the bolt to the washer which was at the end of the thread.

The inspector passed off my sample as A1 with 100%. I was subsequently promoted to an ERA5 Petty Officer and was given a new uniform. A year later I sat for another test and was promoted to an ERA4. Three months later I was selected to serve as Trainee Officer and had to go back to wearing a sailor's uniform with white webbing again.

It was during my Officer training in 1955 when Ian Hamilton and I decided to go to England for an adventure. I had already served six years in the Navy Reserves and hadn't renewed my contract so they couldn't keep me. Upon my resignation I received no recognition of my service. It wasn't until around 2017 that I read in the local paper that the Reservists were entitled to a medal for serving in the Australian Military during the war in Korea. After many letters to the Department of Defence, I did receive a medal and finally felt recognised for my time and great effort as a Reservist. I then joined the RSL here in Queensland.

Me and my friend (Spiks) at Garden Island (1952), after three days aboard the HMS Telemachus British Submarine on loan to RAN for training. My rank: Stoker 1st Class S11200.

## My 21st surprise red MG

Mum and Peter threw a 21st birthday party for me. My guest list included Barbara Mearns, Ian Pilz, George Skinner, Don Potter, Ian Hamilton, Bob Patterson and wife, David Johnson, Gareth Nettherim and Don Humnell, who hitchhiked up from Melbourne for the occasion. I remember feeling somewhat unwell that day, having a tummy ache, but Bob put it down to my excitement. On the morning of my birthday, a friend of Peter turned up in a bright red MG and told me that it was mine. I was thinking, "*Wow*," but he quickly added, *"Till midnight only."* What a letdown. I did enjoy spending the morning driving around Mosman in this flashy car. I drove down to Balmoral and it was here the MG broke down. The gear stick had become stuck, so the owner came over to Balmoral to fix it for me. My girlfriend at that time was Barbara Mearns. She gave me a lovely leather stud box and various books on the Australian Navy for my birthday gifts. Later in life, she married one of my old friends.

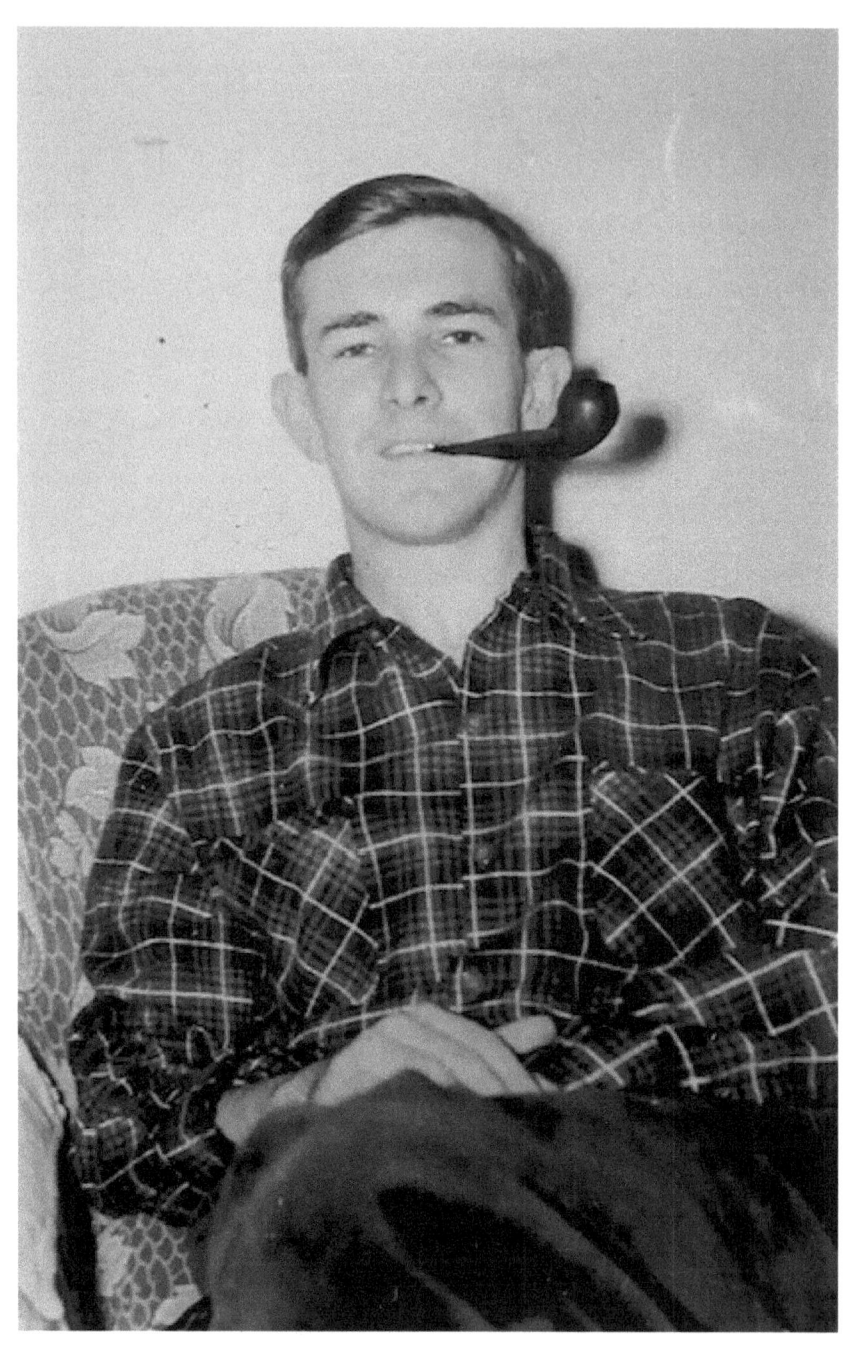
Smoking my pipe feeling confident (1952).

## Shell Oil Company (1953-1954)

My step-father befriended the NSW state manager of the Shell Oil Company in Carrington Street, Sydney, who was also living at Blackheath. I was introduced to this fellow to explain my work background. He said they didn't need anyone in engineering but needed someone to run the company-owned theatrette. I accepted his offer, thinking it would do for now, even though I had no idea what a projector looked like inside. I knew it would be fun to learn something new and it didn't take me long to immerse myself into the new role.

In conjunction with Sydney University, visiting university professors from around the world would hire the theatrette to show new projects or ideas on film. I remember one such film being shown of a gruesome surgical procedure, which put me off quite a bit at the time. One day each week we would have a salesman hiring the venue to showcase their new products. Another day was for pensioners. All these old dilly-dallies came in to watch comedy movies or what have you. The two office girls came down from the office and put on their usherette uniforms. They would sit at the back of the theatre, once everyone was seated. I was in the projection room, with a little window view into the theatre. I would hear from the girls who were seated underneath my window, *"Terry, would you focus the thing."* I would have to get a pair of glasses on so I could get the film sharp. Sometimes the girls would call out to me, *"It's too loud,"* or *"It's too soft."*

### *Henry Litoff's music*

One prominent event happened at the theatrette that has always stuck with me. The English composer, Henry Litoff (1818-1891), wrote a concerto, and his piece is still extremely popular today (you hear it on ABC radio at least fortnightly). In 1954 it was virtually new to the audience who had not heard it before. I used to play old dance music records at the beginning and at interval time slots during the films. As I chatted with the boss one day, I asked him if I could update to classical music instead of the dreary old music. He agreed happily and dug out of his pocket £2, suggesting I go down to Swaines to buy a couple of new records.

At the record store, I spoke to the shop assistant. She mentioned they had just received a new batch of records from England.

She played one for me. The music was absolutely magic to my ears; I felt this immediately, so I bought it. I played this piece at every performance because so many people loved it. I would always get asked what the lovely piano piece was called and where to buy it. I was able to utilise the music notes in sequence with the dimming of the lights - lowering the music volume and then bringing the Queen up on the screen before the film commenced – to perfection. It took me some months to perfect this process. I was so proud of my work.

Music records were made in a 10" or 12" length. The 12" were usually for longer concertos or operas. Popular music was recorded onto a 10" that lasted three and half minutes exactly. I was able to play other music until it was nearing the last four minutes or so, then I prepared for the exact three and half minutes before the scheduled film was due to start, by playing this specific piece of music by Henry Litoff. I never heard that music for decades until the 1990s when it was played a lot again. I believe I may have seeded in some peoples' minds that music of which helped to popularise it today. It's only a short piece.

When the projector was turned on, the numbered countdown timer appeared on the screen, "5-4-3-2-1", at the beginning of every film. I then opened the flap for the film to commence. The projector operated on a carbon light. In other words, two sticks of carbon would come together and spark, throwing a light onto a mirror and then throwing a light onto the 16mm film, which would be fed through the projector and onto the screen. They were automatic machines but if they got out of alignment from not being a perfect distance apart, you had to put the flap down on the projector so the audience couldn't see the adjustment being made. When this happened, I was able to restore the light to its correct position and raise the flap to continue with the show. For films that required two reels, you would get a little flash memento when you had to start the second projector with two flaps. The lights had to be right, and hopefully, there would be no awareness from the audience when you went from one projector to the other.

My boss at the theatrette, John Heyer (1916-2001), drove a Rover motor car. One day he asked me to collect something for him handing me his car keys but I still didn't have my driver's licence. I took his keys thinking, *"I will give it a go."* John was busy making documentary films for Shell. Nearly all the films he made were for motor car driveway performance and maintenance, like checking the oil and getting petrol. He also produced films for sailing safety on Sydney Harbour, or bushwalking tips through the Blue Mountains.

### *Movie film, 'The Back of Beyond' (1954)*

In 1954 John became quite ambitious and produced the film, 'The Back of Beyond'. The hour-long movie was about a man called Tom Kruse, who had an old army blitz wagon a 1936 Leyland Badger mail truck. From Marree to Birdsville, over 500 kilometres one way, once per week, Tom drove with the mail, food stores and fuel supplies up and down the Birdsville track. The filming involved shooting the truck strategically as it drove along the tracks doing its usual errands, of picking up or delivering supplies. The filming would shoot a frame into the back of the truck, where there would be a large can of Shell oil strategically placed for advertising purposes and when unloading at the other end another can of Shell oil would be framed into the filming. The documentary was a clever way of advertising for Shell and never felt like it was in your face like advertising is today.

Upon completion, the film was submitted for exhibition at the Cannes Film Festival in 1954 and it won first prize. My job then was to show the film as often as I could. When I was not doing showings at the theatrette, I would take a truck with a projector and go bush to the Shell gas stations. They would hand out invitations to their customers, placing an advertisement outside promoting the 'Back of Beyond Film'. People would sit on the ground, bring their own deck chairs, or whatever they could find. I would hang the screen on a branch of a nearby gum tree with the projector on the back of the truck. John received a Member of the Order of the British Empire (MBE) in 1955 for his contribution to producing this film. The documentary won many worldwide awards including the prestigious Grand Prix Assoluto, at the Venice Film Festival in 1954.

This time was quite amusing as I still didn't have an official licence to drive a truck to the outback destinations. In those days companies could supply their drivers with licences. The Shell Company driving test for a truck driver was quite difficult to obtain because you were carrying high-octane fuel in the big tankers. The inspector who taught the drivers and handed out the licences took me for a drive to get my license, but I failed. I couldn't restart the truck on a hill. My plea to him was that I only needed to drive a small truck to show the film in the outback, so he asked me to get into his car and drive around the block. I did this with ease and he gave me a Shell licence, warning me not to drive any big trucks.

After I finished showing the film in the bush, I was back in Sydney only a short time when John asked me to get ready to go down to Canberra in a car from the Royal Car Hire company with a personal driver. He said, *"Someone important wants to see the film."* Off I went, having been only given a house address. I was asked to set up the projector on the verandah, with the film being projected through the opened glass doors onto a screen in the house. The driver helped me set up, then went back to sit in the car. I couldn't see who was in the darkened room watching the documentary and they couldn't hear me from my position on the verandah. When the film was finished, former Prime Minister Sir Robert Menzies (1894-1978) walked out onto the verandah to thank me, inviting me inside for a cup of tea. Both Robert and his wife, Dame Pattie Menzies, were charming company. We had vegemite and lettuce sandwiches with our tea. Robert told me he had thoroughly enjoyed the film. It was a big thrill for me to meet Sir Robert Menzies so personally.

## *The call of adventure beckons*

During my time at Shell, Ian and I had often discussed going to England in the future. It was around Christmas time that he phoned me to say he had booked a passage for two, going to England on the Orient Line, aboard the *Otranto* Sailing Ship. Instantly I thought, WOW, this is terrific but I only had about £10 or £12 in my savings and needed to raise the money for the fare swiftly somehow. This felt like a once-in-a-lifetime opportunity and I wasn't going to let it pass me by. It also meant I had to hand in my notice to the Australian Navy Reserves, where I had recently commenced training in their Officer Training School. My step-father, Peter, was not happy with my decision to leave the Navy Reserves mid-way through Officer Training but my decision was firm and I wasn't going to miss this world adventure with Ian.

At this time, I was living with my mother and Peter in the garden shed at Mosman Bay. We discussed my rental and they agreed to keep the shed whilst I was overseas, as I was paying 10 shillings per week. They decided to rent it out for 15 shillings per week as a granny flat until my return.

## *Introduction to High Definition Films in London*

One day a jolly sort of guy came into the Shell theatrette with a booklet on closed circuit television, with a film made by High Definition Films. His company wanted to sell new equipment to the Australian markets that would enable them to record a live program on 35mm film and screen it on television at a later date, which was expected to reach Australia in the 1950s. This salesman gave me the impression it was standard but in conversation, he told me that it was only in the developmental stages. They expected when television had started it would be available. Black and white televisions were introduced into Australia in September 1956, but it was not until October 1974 that colour televisions arrived on our shores.

I told the salesman I was very interested in what he had learnt and wanted to know more. Later in the week, he called me to have a conversation discussing a company called 'The High Definition Films', run and owned by Sir Norman Collins (1907-1982) who was on the British Broadcasting Corporation (BBC) Board. Following the telephone conversation, the salesman said, *"As you are going to England soon, he would seriously consider you for a job."* Delighted, I phoned Ian saying I was set for a job when I arrived in London and we would be on our way shortly.

I had now received my passport and it showed that I was a British Subject and an Australian Citizen. It was signed by the Governor of the day, Sir William Slim (1891-1970), who was Australia's Governor from 1953 to 1960.

Today, some sixty-plus years later, I found the salesman's High Definition Film booklet he gave me back then. I had carried it with me to England hoping to acquire a job with High Definition Films in London. I'm so pleased that I still have the booklet now.

# CHAPTER 6

# World Adventures (1955-1961)

### Voyage to England

The *SS Otranto* was an early passenger ship from the Orient Line doing frequent trips from Sydney to Southampton, taking roughly five weeks with many stopovers along the way. It was a bright and sunny day for my departure from Sydney. Boarding was set by 10am for a noon sailing, so I invited lots of friends from the Mosman Rover Crew to be there. My father came down, along with his friend, Sir William Dobell, who kindly slipped secretly into my pocket a £10 note, only for me to discover it when we were well underway.

SS Otranto Ship (1955).

My step-father arrived with a number of people, including Dr Gilbert. The Mosman Rover crew, about fourteen in all, gave me a very cheery send-off, doing a Hakka on the Wharf as the ship slid off, breaking all the bright streamers. Two girls I knew were also among the guests at my bon voyage gathering. One was my ex-girlfriend, Barbara, from when I was twenty-one. She was crying. I was very touched by her feelings, as I wasn't aware of how she felt about me. "*Ah well, what a shame,*"

I thought to myself. My voyage was soon underway, the ship escorted out of Sydney Harbour by the *Captain Cook* pilot boat. Coincidentally, the same pilot boat had escorted my dad and I when we were being relocated to Perth during the war, in 1942.

On our second day we stopped in the port of Melbourne, where I was to pick up Ian. My best friend, Ian, was a junior copywriter. I could see advertising was in his blood and he would pursue a career doing what he loved. He was also an up-and-coming novelist. Ian got started with short stories for the Women's Weekly.

We arrived about midday and Ian was there on the dock to welcome me. We dashed off in a borrowed van to his place for his somewhat drunken farewell party. He had many friends in Melbourne, from his school days and people he worked for.

We did celebrate a bit hard, going on till the early hours of our departure day, with very little memory of how the hell we ever got back to the ship on time. Like me, he had loads of friends and family to see him off, including one special girl, Noelle. I had been befriended her a year or so ago and I wrote to her often. Assuming that she was my girl, I decided to propose to her on the deck in a light-hearted way. She didn't refuse but also wouldn't accept. She promised to wait for my return to Australia in the future. That was dashed because shortly after I arrived in England, a letter informed me and Ian that she had become engaged to a wonderful man from the Royal Australian Air Force. We did have a quick meeting in Sydney some twenty years later but now I have lost all contact with her.

Our next stop was Fremantle, Western Australia, where I had loads of relatives to call upon. I hadn't seen them since 1945 when Mum and I moved back to Sydney after the war. George Winning, an old school friend from Wesley College, is my closest friend in Western Australia. We have continued to stay in touch with each other. George bought himself a Piper Plane after getting his pilot's licence. He insisted on giving me and Ian a sight-seeing tour over Perth's city suburbs, especially around the South Perth area where we grew up. Then, back on the ground, we enjoyed a quick call on my aunties; Clal, Marjore and Dora. We hastily sighted our old school, the old house in Darley Street and of course the old mill along with George's old house, then it was back to Fremantle to be on our way again. I felt quite dejected as many of my Perth family, both on my father's and mother's side, didn't come down to hold streamers as we sailed off.

Piper plane ride around Perth for Ian and me,
with pilot George Winning (1955).

Ceylon, now called Sri Lanka, was four days of sailing from Fremantle. We passed the sister ship, *RMS Orcades*, crossing and slowing down at the Coco (Keeling) Islands in the Pacific Ocean, northwest of Perth. We later arrived in Colombo, the capital of Sri Lanka.

As we ambled through the market area known as the Pita, a young man grabbed Ian by the arm and told us he had stolen a diamond ring and didn't want to be caught with it. He showed it to us and it looked impressive in the gold market. It was in a small blue velvet case. He would accept £10. Ian jumped at the chance to produce the money and recovered the jewellery box, putting it quickly into his backpack. He was so delighted with his purchase. It was not long before he retrieved it from his backpack to look once again at his treasure-buy. Our young friend in the markets had done a swifty on Ian. With a sleight of his hand, he had switched the blue velvet case with another. Ian now had a coloured-glass brass ring. Once bitten twice shy, I won't repeat the language Ian used, but it would have curdled cream, maybe a shattered church window, or something like that.

After joining up with a number of friends from the ship, we met at St Riviera for lunch, where the girls bought good quantities of moonstones that they later on sold in London for a handsome price. Upon returning to the ship, we met a local family who were boarding our ship to go to France. Ian fell for the teenage daughter, her surname being De Silva. Ian reunited with that family many times in the future.

Further, by strange coincidence, as I lived on Bruny Island during my retirement years, her brother Melvyn De Silva became my next-door neighbour. He was a radiologist at the

Hobart Public Hospital. When his sister died, Melvyn went back to Ceylon for her funeral. He had come down to Bruny Island with another doctor, Dr Ian Johnson, who he worked with in Sydney. Dr Johnson had written quite a few books. I have one. He translated volumes and volumes of Chinese poetry into English. The Chinese were honoured by his incredible work and awarded him some sort of recognition for his enormous efforts.

During our voyage, we had many occasions when Ian would ask me to entertain Melvyn, so he could have time to spend with Melvyn's sister. After a couple of days in Ceylon we were off again. At the mouth of the Red Sea, we sailed into the Suez Canal.

As we approached Port Said, we booked in to cross into Egypt via the pyramids, but as we approached a battle had broken out in Cairo. A British soldier was shot and killed, and as a result, all trips to the pyramids were called off. So, we decided to visit a nearby city instead. Leaving the other passengers to explore, Ian and I went in a taxi to an old town.

I was on the lookout for a camera shop as I wanted to acquire a 35mm film camera, as I didn't have anything like that. I nearly bought one in Sydney, but it was going to cost me £150. When we arrived, it was devoid of tourists, so we had the camera shop to ourselves. We were invited in to squat down on a carpet and drink a very strong but delicious coffee before we started shop talk. The camera I wanted was a German brand called 'Diax', not the luxury camera 'Leica' that I would have preferred. The shopkeeper offered to sell the Diax for £100. Apologising, I told him I didn't have much money and asked if he had anything else.

He replied, *"Well you like this camera, will you accept £50?"* It was still too much. *"Will you accept £25?"* I apologised for the trouble and explained I only had £10, being a gift from Bill. *"Yes,"* he said, *"I'll accept £10 and throw in a flash gun to boot for another £1, and a Ronson Proof lighter because you smoke a pipe."* I accepted his offer. Today I still have that lighter and the camera. Ian had the same fun and finished with a Lietz 8mm movie camera, also a Ronson lighter. Boy, we did do well that day.

We headed back to the ship for the long passage across the Mediterranean. Arriving in the port of Naples and after a long look around from the deck, with hands in our pockets, we finally disembarked and explored the town. The next morning a bus trip was offered to have lunch in Amalfi, a small hillside fishing town south of Naples. We dined on a terrace overlooking the bay and guess what...? Spaghetti bolognese was the only fare considered. It was the best I had ever tasted, and even to this day, has never been surpassed.

Along with our little ship family were girls from the Sydney Symphony Orchestra and a beautiful young Jewish girl, Aviva Cantor, who was on her way to Paris to study French. I found her sitting on the bus and sat next to her from Amalfi to the ruins of Pompeii, some 36km away. We were very attracted to each other and became constant companions in Italy. Pompeii was a sheer delight to both of us and we no doubt were very interested in the ruins but in each other as well. At the end of the tour, we were roused from a discreet corner to hear a voice calling out, *"Please get back on the bus, as we are late and other passengers are getting restless."* That night we all went to a nightclub called the 'Petit Noir Chat' just out of Naples for a fantastic evening of eating and drinking. The following day we visited churches and museums and finished up at the same club as the previous night. During the evening someone noted the time, discovering our ship was due to set sail in about 20 minutes. Quickly finding a cab and pressing the driver to hurry, we had a nervous trip down the mountain, through the old town's winding roads, and onto the pier as we saw the gangplank come down again for our group. We crawled back onboard but farewelled Aviva, as she planned to catch a train to Paris. We made plans to meet up again but it was not until 1995 that I met her in Double Bay, Sydney, to launch her book, 'Jewish Women Jewish Men: The Legacy of Patriarchy in Jewish Life'. I still have a copy of this book. By then she was happily married with a grown-up child and her husband was a poet, of which I have one of his books too.

The next stop was the British garrison town of Gibraltar. Apart from the huge Naval Base, there was only a small township to explore. We had missed out on a bus trip to see the Barbury apes. There was a myth that should the apes ever leave Gibraltar then so would the British. You can imagine the British took great care of the apes. Aboard again, our next port was Southampton, England, where our voyage would soon be over.

With a letter to Sir Norman Collin, the head of High Definition Films and a Director of the BBC, I was confident in getting early employment. By this time, I had spent nearly all my money and so had Ian, who had no job prospects. I also carried a letter from my step-father to his old wartime friend Bud Flanagan (1896-1968). Bud and his professional partner, Chesney Allen

(1894-1982), were a successful duo of English comedians. I was sure Bud would be of help to Ian because of his possible connections in London.

As we entered the English Channel, we noticed a crew member was collecting glasses from the deck from drinking passengers. He disposed of the glasses by tossing them over the side of the ship. On enquiring why: *"Well the ship will be scrapped at the end of this trip so why bother."* So we decided to help ourselves to a couple of ashtrays, we had our souvenirs from the voyage, and there we were disembarking at Southampton. After five weeks, it was wonderful to have finally made it.

Ian's dad had bought a night at a Kensington hotel so at least we would have somewhere to stay while looking for more permanent digs. We boarded a train to Kings Cross station. We had no clue as to how or why we took a taxi, in a famous London cab, to the Kensington hotel, it just seemed the obvious way to travel there.

Arriving at the hotel we checked in to our room, then took off, walking everywhere and exploring London. We soon found Buckingham Palace. It felt like a story book venture.

We watched the Changing of the Guards, then just moseyed around. We sat on the Grand Fountain watching traffic, and people moving, and fantasised where they were going, who they were or where they came from.

Ian and I were so moved by all this, not only by the fact that we were where we were but because we had finally made it... Yes, now on the other side of the world, a long way from where we had spent all our lives. Now we were ready and willing to start a new life here. Eventually, we found our way back to our hotel and had a grand meal in a local pub. After dinner we collapsed into bed for our first night in London.

The following morning, we were at the check-in desk to enquire if we could stay another week. We were informed another £10 would be for a week. "*Wow,*" I said. £10 for a week was going to be terribly expensive. I was curtly reminded that the £10 was booked for one night only. My next question: "*Do you know of any cheaper accommodation nearby?*" We were recommended the Earls Court area, as many Australian tourists and backpackers headed there. This we did but many had no vacancy signs. "*Keep looking,*" said Ian.

At the end of the road, a long brown stone building, four floors up, offered accommodation at 50 shillings a room. We took it immediately. The flat was a large attic on the top floor with a shared bathroom. A penny in the slot for gas heating with a stove but no music after 10pm. The front door of the building was always locked at midnight. There were two beds, a pine table, two chairs, one fry pan, 2 plates, 2 glasses, 2 knives and 2 forks, and a bucket of coal for the fireplace.

We moved in and as we had the day to ourselves, I said to Ian that I would venture to the BBC hoping to meet Sir Norman Collins, though not knowing where I would find him. Ian said that was fine and went out by himself. He found a second-hand shop and bought a radio and some books for 2 shillings and settled in. I promised we would go and see Bud Flanagan the following day.

# British Broadcasting Corporation (BBC)

My trip to the BBC resulted in a lot of walking, which nicely coincided with plenty of sightseeing along the way. I entered a grand building, the home of British Radio and Television. I felt lost in this huge building but finally found his office. I nervously waited outside. A casually dressed man with no tie, Sir Norman Collins, invited me into his office. He turned out to be a delightful man.

I gave him my letter of introduction from the Sydney salesman. Sir Norman was impressed with the letter of introduction and the information contained within. Unfortunately, I never read this letter and was sorry I never opened it but I got what I wanted and that was an interview for a job. *"Fine,"* he said and made a telephone call to a Dr Spooner of Highbury Studios. An appointment was made for me to see Dr Spooner on Monday morning at 9am. This being a Friday, I had two days to plan for it.

Thanking him, I hurried back to Earls Court to tell Ian. I wanted to encourage Ian to call in to see Bud Flanagan at the London Palladium Theatre that night. Bud was appearing in a show called 'The Big Crazy Gang'.

I had no trouble in finding the theatre as the show was very popular in London. Bud always appeared on stage with a battered straw hat with the front turned up, making it identifiable. We went straight around to the stage door and were let in but told to wait until the interval when Bud would return to his dressing room.

It was very exciting for me to be there, being behind the stage during the show and having much infatuation in meeting Bud Flanagan in person. There was a lot of activity, so many people were coming and going backstage. Showgirls in very skimpy dresses, stagehands and other actors busy going about their business. Our waiting time went very quickly. The doorman had a brief word with Bud and next thing, there we were, in his dressing room with that battered straw hat on the hook behind the door.

He read the letter and said, *"Yes, I have fond memories of your step-father Peter and the war when we were neighbours. We saw each other frequently."* With my background, he couldn't help me, not knowing anything yet about television or having other connections to the electronics industry, but he stated, *"Ah yes,*

*I can help Ian. I have good friends with Peacock Advertising and will follow up with some enquiries."* We both turned to leave, then I asked Bud if I could just wear his straw hat for a moment. He happily gestured to it and said, *"Yes, by all means."* I tried it on for one very exciting moment before we left.

# High Definition Films

Well, Monday morning arrived and the fun began. I was escorted promptly to Dr A. M. Spooner's office, at their studios in Highbury Park. He said he had a very favourable call from Sir Norman Collins and would be delighted to give me a job, but I had to fill in an application first. No problem, except one small question: 'name of union and number'. I explained I had only been in England for less than a week and I didn't have a chance of becoming a union member. *"No problem,"* he said, *"take this form to the Union Office and get a membership."* The Union Office happened to be just over the other side of London. Using the last of my money to get there and back I found the office, surrendered my note and a gentleman came out of an office to tell me, *"No union membership until you have a job, goodbye."* So I went back to Dr Spooner's office, on my last penny, with the news. He said, *"Sorry, but I can't give you a job until you are a member of a union. However, there may be a solution here. I can employ you as a temporary cleaner, then you can join the cleaner's union. When you get that I can promote you to a technician, so you can then join the other union that covers all departments in the television trade."* I agreed and Dr Spooner said I would have to receive a cleaner's pay for a bit. I asked him if I could have an advance, as by now I had no money in the world. *"Why certainly, but I shall have to loan it to you myself as you have only just got the job – starting tomorrow at 7.30am."* I went back to our little room and told Ian, so we decided to celebrate with dinner and a beer at the local. What could have been more wonderful? I felt thrilled. Ian also soon landed a good job working for Peacock Advertising.

### *The New World of Electronics*

At High Definition Films (HDF), there was a whole new world of learning for me. This was my initial introduction to the electronic world. My job was not cleaning, this of course was done by contract cleaners. I was introduced to a guy in the workshop, along with two girls in the front office, one being a typist and one a receptionist. Dr Spooner had his own office and there was a vacant office next to his. He introduced me to the store's person and two people in a laboratory, where film was processed. I inquired to a fellow worker what he was doing, he replied, *"I'm making up a radio circuit but firstly doing the drawing to plan it."*

My initial job was to take things apart by disconnecting the various components with a soldering iron, breaking the connections then saving all the bits and pieces that could be used later for another project. A fellow worker was showing me, from the drawings, how to interpret each component that went together and why it worked. I thought to myself, 'It's magic' because it was something I had not seen before. Back then I had a little blue book, which I haven't found yet but believe I still have, where I took notes. My book showed me the colour coding and the resistors, which had number coding accordingly, so I knew what different valves looked like and what they were used for. There was a drawing of a transformer and what it was used for, showing all the values and what have you. Being something new and of great interest I could visualise things for this job very quickly. We were making this equipment by experimenting and manufacturing at the same time. The components were fitted to a board with a frame of angle iron. We put the strips of metal onto the front with the controls, and on the back we placed the components. We were able to set up a bank of this equipment to work on and started manufacturing them.

We had a darkroom with technical personnel operating equipment that processed an electronic signal from The Finsbury Park Empire Theatre next door. It was an old picture house, now a television theatre with a stage. Sometimes there was an audience but always a TV camera. They mainly produced interviews that were broadcast live, as there was no means to record the program. Hence, the HDF company was a busy building. We created the necessary equipment to record the interviews on 35mm film that could then be broadcast again. The system already built worked but was very large and bulky in size, taking up a whole room just for the electronic equipment. Our job was to redefine the equipment, reducing it to a working site. This was back in 1955. Today a mobile phone will do the same job and more in a fraction of the size.

Back then, there were radio tubes and transformers taking up a lot of space. When the transistor came along all that disappeared. It was a remarkable change to witness.

At the time we had a piece of board we could nail panels onto, and solder wires to various components. With a meter and a pair of headphones, and a probe, we could see what was happening to this magic signal from the cameras next door.

As time went on my boss would draw out a circuit, and with my soldering iron and pliers, I put together parts that previously I had removed, possibly even adding some new ones. He would test it with his meter probe and headphones, check my creation, and then point out some changes to make it work better. It might have seemed that's about all there was but in reality, there was much more going on. We had a cathode-ray oscilloscope, which measured current variations on a screen. We would then take a photo of the image. The negative was taken into the darkroom in the other lab for developing. I could process the negative into a black and white print for all to see how our progress was being made on the equipment: parts that were being built much smaller than the original ones.

Now we were developing these boards into smaller panels that could be screwed onto a frame, connected to each other by cables running down the frame at the back. One day I was receiving a bunch of these panels from the local manufacturer and noted the delivery docket of what we were paying for them: £2 each, far too much at that time. "*Wow.*" I marched into Dr Spooner's office to announce the greedy discovery. "*Why can't we make our own panels and save a lot of money?*" Dr Spooner said they didn't make them as no one knew how to do it. When I told him that I could, he inquired how I knew. I explained to him of my engineering background, so he gave me £10, asking me to go ahead and buy the tools. I went to a second-hand yard buying pre-war tools, a guillotine for £5, and a few other things. The spare office was used mainly for unwanted equipment, so I took over this room and created a metal-working workshop.

I drilled the holes for components, sprayed the panels white and hung them to dry. This project ended up costing less than half of what we were purchasing them for originally. They were thrilled with the outcome. Dr Spooner said, *"You amaze me with your innovative mind and ability to bring new ideas to life."* I was promoted to Technical Manager with a big pay rise. I had one member of staff to help me, of which I do admit, I enjoyed giving orders to. I often used wood for the panel testing. For one particular piece I polished it up with linseed. The wood came up so well, that it was far too nice for the panels, so I gave it to Dr Spooner as a chopping board for his wife. I received a delightful letter from her that she used it every day.

My job at HDF was not entirely making panels. When the testing was done in the main lab there was a monitor, like a television set, that projected various patterns we had made. It was very important to keep a record of these patterns, so I took a photograph of the patterns and then processed the film. I learnt how to develop those films into prints. I thoroughly enjoyed doing this. Generally, I had a great time working there. The workers were the funniest guys, all Goon Show fanatics who would speak in 'Goon language' with plenty of Goon jokes. It was always a happy, joyful atmosphere.

One of the lads from HDF, Johnny Mattison, a Teddy Boy, offered me and Ian a super flat at Crouch End in North London since they were looking for a couple of tenants. By then we had only been at Earls Court for a week or so. However, we decided to go and have a look. Ian and I met Mrs Medway at the flat in North London. She was a very nice Jewish Mumma who thought we were lovely and hoped we liked the rooms. The flat was on the 3rd floor with a bathroom on the 2nd floor. It had a kitchen, a spare room and a small bedroom with the tiniest fireplace. There were two beds. The rent was about five shillings a week and we decided there and then to take it. Johnny had a car so he gave us a lift back to Earls Court to collect our bags and say goodbye to our neighbours, a couple of lusty Scottish girls that we had started to show a bit of interest in – but that soon passed when we left.

Sid, Mrs Medway's husband, had a stall at Petticoat Lane every Saturday morning. He sold very thin aluminium saucepans and asked us if we would like to go too. What fun for the first few weeks! We would just stroll around, but then Ian got a good job at the markets selling second-hand suits for £1, which was quite a lot of money then. I think this job lasted only a week or two. I went around and ridiculed the rubbish that was being marketed till I was severely threatened, so we didn't go to the markets anymore.

After Ian and I had been in London for about six months, we started to go to the Down Under Club. A club for mostly Australians, apart from a few British who were friends of the Aussies. We would have a pint of 8% alcohol rough cider and get a pack of 10 'Players Weights' cigarettes for our night out. At the Down Under Club I met Lou Grey, a theatrical agent for the stars in England who became an acquaintance of mine. I met many interesting people in London.

### *The Spy Catcher*

I'd been at HDF for about two years when a new member joined the executive staff, Peter Wright. We were never formally introduced but over his time there, Peter showed a close interest in all that was going on and was given his own office. He spent a lot of time talking with each person, individually. As a group, we speculated that his interest lay in the fact that a TV camera operates silently and the receiver of the signal from the TV camera would be recorded for later viewing. This technology was nothing new, for that's exactly what we were doing. So why was his approach so different we thought? We surmised he was from MI5, the British Intelligent Agency, foremost a spy organisation and what an asset he would be in there, intelligence gathering. Today of course it's as common as an iPhone or a CCTV, especially for road safety and surveillance. Peter turned up for my wedding and I sort of wondered, what is this bloke doing?

In the late 1980s, an event gave credence to this story. An MI5 agent retired and explored a lot of the secrets behind MI5, only to then publish his findings in a book titled 'The Spy Catcher'. It revealed that MI5 were looking for a technology to film through a glass wall whilst a prisoner was being interrogated. That was what Peter Wright was learning at HDF. When the book was published Peter was swiftly sued by the British Government for giving away state secrets and there was a big trial held in Sydney. The young lawyer defending him was Malcolm Turnbull, later to become an Australian Prime Minister. I enjoyed going into town during the proceedings, watching the 'Spy Catcher' author, Peter, walk into the court room, as I knew him from HDF. I always wished him good luck and his lawyer, Turnbull, put on such a performance in court he successfully won Peter's case much to the dismay of the British Government.

Many, many years later when I was living on Bruny Island, I found out that Peter Wright was living on a horse stud in Tasmania and was known to drink at the local pub. My doctor friend, Dick Geeves, knew the Spy Catcher's daughter and asked me if I wanted to meet the man, who lived not far from me. I visited the pub Peter frequented, asking the barmaid if she had seen him. I was informed that he had just left, so I walked outside, looking up and down the road, but I couldn't see him anywhere. I did find his house and knocked on the door. Mrs Wright answered, saying Peter was not there. I asked if she could tell me anything about him or the MI5 recording equipment at HDF in London, as I had worked with him at HDF. She was unable to say anything about what was written in his book relating to HDF and sadly, it was not long after, I heard that Peter had died of old age. It was a strange coincidence we ended up so nearby again, and I'm disappointed still that I never got to see him again and have my questions answered.

### *Colour production*

One day a fellow suggested to me, *"Wouldn't it be fun to produce a bit of colour on the screen?"* I said, *"You can't,"* and his instant reply was, *"You can."* He flashed some light on the screen, making a rainbow and saying, *"There, you can do it."* It was a matter of presenting this kind of effect on a TV screen. He further explained, *"I've created a drawing of a Christmas tree and in each frame the tree is off-centre. We hope the Christmas tree will light up in colour but we need to run it through a projector at various speeds, however we haven't got the equipment yet. I want you to reproduce the film in a lengthy quantity and put it through a projector, at different speeds."* At the time, recording film on 35ml was to be used later and worked beautifully. In my little blue book is a strip of film of the said Christmas tree image from that day.

This fellow visiting HDF was Sir Barnes Wallis (1887-1979), an engineer who had invented the skip bombs for WWII. The film 'The Dam Busters' was produced in 1955 retelling the true story of Operation Chastise, using those skip bombs to take out German dams in 1943. He frequently visited HDF and we soon became good friends.

# My first film production on 35mm

Not long before I left HDF, Dr Spooner came into the workshop announcing that we had completed an entire chain of works. This meant we had a fully working new system including a camera power supply, and the ability to record and print onto 35mm film. It was all ready to distribute and broadcast previously recorded events. Now we were excited to do our first test run. Thinking ahead I asked Dr Spooner if we could use a script and maybe introduce a very short play. He agreed but wanted no more than five minutes and not in full colour. Our test run was set for 10am the following morning, as he would like to invite Sir Norman and a few of his colleagues from the BBC. I wrote a very short story of a few words, with a little action. It went something like this and oh yes, I elected myself to be the cameraman and producer of this short film. I needed one of the guys to operate the suspended microphone and another to handle the cable trailing the camera. It went something like this...

**Shoot 1 – The camera opens up, viewing the door to the secretary's office and a message boy (office boy) knocks on the door and goes in.**

Shoot 2 – The camera moves to the doorway to see the message boy handing an envelope to the secretary at her desk. She opens the letter and reads to herself, then a big smile comes to her face and she runs out and moves through the door.

Shoot 3 – The camera moves back into the workshop, to see the secretary come through her office door, down a passageway and into a workshop. Then across the floor to Dr Spooner's office, she knocks on the door and goes in. Instantly he rises from his desk and walks into the workshop with the secretary behind him. He stops and asks the personnel to listen to the good news. The equipment we have been working on all year is now complete and ready for a first test. His audience cheers and claps. The news and film fade into black.

At the end of the filming, I took the negatives into the dark room and printed them ready for viewing. Unfortunately, the film was taken away and I never got to see it. It was viewed by the BBC officials who were delighted with the result and this became my first film production.

Me the Pye TV Camera Operator at Highbury Studios, High Definition Films (1955).

# Jacquie

One night, a girl we knew from our trip on the Otranto ship invited myself and Ian to a party coming up. Not wanting to miss out, we went along – and what a night it was. Drinking lots of cider and meeting loads of 'her people', I got very friendly with a New Zealand girl called Jacqueline Crum wearing big funny glasses. She seemed such a happy-go-lucky girl, so I quickly cottoned on to her attention. She was working as a Karitane (a baby/child care nurse), looking after a newborn girl. The baby's mother was a performing opera singer, quite well known and the father was a ship's Captain in the Royal Navy and was away a lot. Jacqueline and I planned to meet up again after the party. She was attractive, the same age as me and always laughing a lot. Jacquie was generally good to be with, I thought, so we met up and courted for about six months until I finally popped the question, which she accepted.

### *Highland Adventure*

Two of the girls from the Sydney Symphony Orchestra, a cellist and bassist, asked if would I like to drive them on a week's holiday to Scotland to see the Edinburgh Festival. Oh yes, I was only too happy to oblige. I had been working for about a year and was now entitled to a week's holiday. Leaving London and going north, firstly through the Lake District, we travelled through many little villages. They were so charming that it was always a pleasure to stop and look around. I'll admit to going to the 'Gorbals', a district in Glasgow known for its poverty and overcrowded dwellings, and getting very drunk with some fantastic Scots. Then we headed to Edinburgh for the festival which was great fun.

The following day we continued north to John O'Groats, the most northern part of Scotland. We bought some smoked haddock and a few potatoes. We stopped by the roadside, next to a wee stream to boil the fish and potatoes in the same pot. With a bottle of scotch and the magnificent fresh water from the stream, seated upon the fields of heath, we dined. A meal never to be surpassed in all the meals I have had since then, having had quite a few; it was magnificent. There was a bit of lovemaking and the world was mine. What a day!

My brief holiday to Edinburgh had no bearing on my rising relationship with Jacquie. We had a few weekends away and, well, I guess we simply fell in love. Jacquie told me that she might be pregnant and we should do the right thing. Neither of us wanted a wedding, so it was decided that a Registry Office marriage would do the trick. She advised me that she would not have a baby under the British Health System and was certainly not interested in going home to New Zealand just yet. Jacquie liked the idea of migrating to the United States of America.

I applied to the USA by putting my name down on the list at the American Embassy. The officer at the Embassy, however, informed me that they were only taking limited numbers from Australia. Specifically, they wanted only the citizens from countries who had represented the USA during the American Civil War. Because Australia was not represented by many people, it would be a little while before my name would come to the top of the list for a green card (which it didn't, until 15 years later). I asked the officer, *"How about Canada?"* and he replied, *"Yes, Canada will be fine,"* so we applied and were accepted promptly. This is how it was decided: *Canada here we come.*

Jacquie's dad sent a cheque for £10 to pay for our wedding reception. I asked Dr Spooner if he would let me have Friday off to get married. *"Well of course,"* he replied. I went on to say, *"We would like you to come to the reception."* His reply was, *"Well of course, that would require..."* but I interrupted, continuing, *"We would like the entire staff of HDF to come actually."* He kindly responded with, *"Ok, I'll just have to declare a half day holiday on behalf of everyone, thank you."* I handed in my resignation to take place a week later, upon the return from our honeymoon.

Norman used to come down to the workshop every so often, wearing daggy old shorts and sandshoes, but he drove a beautiful Rolls Royce. He was mainly an author, writing a lot of books about England. 'London Belongs to Me' was published in 1945 and 'Annie' was published in 1942. I had copies of both but I think they had to stay down in Tassie along with most of my book collection. He was later appointed a knighthood by the Queen. He was so easy-going, such a nice bloke. Norman kindly gave me a letter of referral addressed to the manager of the Canadian Broadcasting Association (CBC).

Jacquie and I hadn't planned on a honeymoon to anywhere specific until one day we were crossing Piccadilly Circus and bumped into an old friend from the 1st Mosman Rover Crew, Mike Smith. Apart from our small talk, I invited him to the reception. He replied, *"Sorry I am going to drive to Paris that afternoon."* I thought quickly of a Paris honeymoon and said, *"Yes we can skip the reception and meet you at the given time and place."*

### Wedding Day (1956)

During our marriage ceremony, Jacquie had some contact lenses prescribed because she thought her glasses held her back a bit but when one of her lenses fell out we all scrambled over the floor to find it. Bea Audet, Ian's new girlfriend, burst out laughing, and then we all did. She was very pretty and hoped that they too would marry soon, which they did. I said, *"Forget the contact lens, let's get out the glasses and get on with the wedding for the ceremony to proceed."* My cousin Bill was there with his trusty camera taking all the photos, so we could see what we missed at the reception as we travelled down to Paris. We did attend our reception for a few moments, mainly to receive and give a few speeches. Jacquie's father's gift of £10 soon ran out, so HDF bought more grog to keep the party going. My appreciation for Dr Spooner holds no bounds; everybody had a great time. The reception went on till quite late that night, even without us.

My Mosman friend rang to say his girlfriend, a ballet dancer living in Paris, had booked us into a hotel near the Bois de Boulogne in the heart of Paris. She was an old girlfriend of mine from Mosman and had remembered me. It was a fast drive down to Southampton crowded into a little MG sports car. We caught the channel ferry to France and then continued our drive into Paris.

He promptly dropped us off at the hotel and simply left. We found there had been no booking made, and for the dinner date that night we had arranged, they never showed up to join us. I've never seen him since and I often wonder what actually happened. Down the road was a wine shop that had a sign, 'room to let'. We booked in for a week with breakfast thrown in. All checked in and ready for our honeymoon in Paris. A wonderful French man and his wife looked after us so well, they were also thrilled that we had just got married. We spent a lot of time in our room, only venturing out in the afternoon to go sightseeing. I wanted so much to climb the Eiffel Tower and see the fabulous view from atop. Every day was warm and sunny but not the day we chose for the tower, which was grey and cloudy. By the time we got to the top, we didn't stay long, going back down to have a glass of wine or two at the restaurant. The Louvre Museum was on my list too. I had impressed on my dad that I would go and look at the original Van Gogh paintings. I loved seeing these, and the marble statues, in particular 'The Winged Victory of Samothrace' and 'Psyche Revived by Cupid's Kiss'. We both enjoyed visiting the Notre-Dame Cathedral, nearby, for its wondrous architecture. Another day out we strolled through the 2088 acres of parkland at Bois de Boulogne. This land was given to the city of Paris by Napoleon III in 1852, to be created into parkland for the city dwellers to enjoy. We watched many young children find amusement in the park, and there was a zoo, market stalls and a play centre. It was just nice to sit by the pond chatting with Jacquie about the present time and what the future held for us both. We were very much in love.

I bought a lot of Gaulioses cigarettes to bring back to England with me as they were very cheap in Paris, being of French origin. Our journey back to London included a train to Le Havre, a ferry to Southampton and another train to London. We only had a couple of days left in London to say goodbye to all before our swift move to Canada. We received a wonderful present from the lads at HDF, a self-made record player for the new LP records.

# Voyage to Canada (1956)

We travelled down to the London docks to board *The Arosa Star Ship* for our voyage to Quebec City, Canada. Our journey across the Atlantic Ocean was short but very enjoyable for us both. We met many people who remained good friends the whole time we lived in Canada. The only amusing thing I remember on board was that the dining room kitchen prepared only German style cooking. To Jacquie and I, it was fine; we liked it very much. But a number of British migrants wanted sausage and mash or fish and chips. The menu was changed to suit the British meal description, but the dishes remained the same, good old German style food.

We set foot in Quebec on the 30th July 1956. Upon our arrival, I asked Jacquie, *"Do you want to live in Montreal or Toronto? Montreal is very French being Canada's biggest city."* She replied, *"Yes I would love to learn French,"* so together we chose Montreal.

In Montreal I headed directly to the Canadian Broadcasting Corporation (CBC) looking for work. Upon meeting the manager of CBC, I showed him my letter of reference. He said, *"How's your French?"* and paddled away in the French language to me, then asked, *"Did you understand any of that?"* My despondent reply was, *"I didn't have a clue"*. He said, *"I'm sorry, you have to have a good knowledge of the French language to work here, however they are looking for people down in Toronto, so I suggest you go there. If you learn your French, you can come back here."* I accepted his advice and we got on a train to Toronto. Arriving at the CBC in Toronto, I handed over my letter of introduction with the wonderful response, *"Boy, do we need you, you can start in the maintenance department immediately."*

After securing the job we went searching for accommodation by looking in the Toronto Times newspaper. We found an ad that seemed available in a suburb just north of the city, so we took a taxi there. We were welcomed by a wonderful lady who introduced herself as Blossom, and blossom she was. Her husband, Yarney, was a cab driver. They offered us a basement unit not quite below ground level. The bedroom, dining room and kitchen were combined into one large open living space with high windows. There was a store room next to the big oil heater which kept the home deliciously warm and cosy throughout the winter. I later turned this area into a workshop, where I repaired TVs. We moved in later that day and then found out we had to share a bathroom with them. Although it was an inconvenience we settled in very nicely and they turned out to be a wonderful couple.

Yarney also had a business on the side called 'Bumpa-tell'. He installed metal sign panels that were bolted to the rear bumper of a car. All cars had back and front bumpers in those days to prevent damage if you got too close to another driver. These signs were used by cab drivers or commercial vehicles to advertise local businesses and I spent my weekends fitting these. Yarney paid me well for the work as he earned a good rental on the signs.

Blossom's father, who lived just a couple of streets away, drove a colourful Cadillac. One day he came around and asked me if could I fix the radio aerial as it wouldn't go up or down. No trouble, I found it was just dirty and needed cleaning. A coat of oil and it was working beautifully. We became good friends. Down the track he showed me his hobby that he loved. His basement was fully converted into model railway tracks, with all the inclusions. He was a carnival or circus organiser, so he had agencies all over the United States. Blossom's father swears he gave Mickey Rooney (1920-2014) his first job. Mickey was a Hollywood star known not only for his acting roles but for his many wives.

# Canadian Broadcasting Company (CBC)

My first job with the CBC was in the maintenance department where we kept all the electronics going. This meant not only what was on the studio floor but in the control rooms. In the master control-room we had projection units, where film was being broadcast. This was mostly for daytime television, whereas in the evening, news programs and other forms of entertainment programs were broadcast. One program that was a presentation on the current hit times, as I remember for a long time, was listening to 'Goodbye little yellow bird'. We had quiz shows, plays and mysteries. Meanwhile, we were popping in and out of the studio control rooms checking that all was well. These were all live shows.

Here I repaired cameras and other technical equipment as a film projectionist. During this time, I had been listening to a lot of complaints about the condition of the studio equipment, for example, camera pedestals, panning heads and sound booms. So, I volunteered to look into these matters as no one else seemed to want to. These were constant annoyances. Well not to me – I just loved getting into these pieces and pulling them apart so any opportunity when this problem came up I was sought after.

## *Squeaky camera pedestals*

One item that was a constant reminder and one I soon found out to be a menace to repair by the other maintenance workers, was that an unwanted sound might appear from a squeaky camera pedestal on the camera panning. Inside a camera, there are a lot of moving parts. If there was a noise on the set, I would have to pull the camera to pieces, literally, sometimes lubricating each part to make sure all were working smoothly.

My work developed further into the maintenance of the panning heads, of which the cameras sit on, the tripods or pedestals, the camera crane, the microphone boom, and that sort of thing. I set up my workshop in the hallway, outside the maintenance office, driving everyone mad. After about a year, they finally gave me an area in the warehouse in a storeroom, where I could set up my workstation appropriately. I took two of the maintenance people with me, becoming a boss.

# My son, Rolf (1957)

As Jacquie waited for the birth of our first child, she continued working in a toy factory making dolls' clothes. She enjoyed this and was very good at it. One night the birth became imminent so we rang our landlord. Yarney was happy to drive us to the Toronto General Hospital. Jacquie's lovely young doctor, Chris Yankou, said *"Your wife has a very small pelvis and might have trouble giving birth, but there is a new product on the market called Xylocaine. If I inject it into her spinal column, it will relax her muscles so she will not feel any pain."* She had the injection and everything was perfectly fine with the doctor announcing, *"The perfect baby"*. Rolf was born a Canadian citizen. Xylocaine is still used for pain relief in childbirth today, some sixty years later.

When I informed my colleagues at the CBC of the birth, they were more than happy to drive me up in a car to the hospital to visit Jacquie, and then bring me back to work. On the day Jacquie could come home, the company ordered a limousine for us. What a wonderful start in life for Rolf! For the rest of our time in Canada, Jacquie was always there to look after Rolf. Our landlady, Blossom, had recently given birth to a son also, so the two babies spent a lot of time together. We asked Garth

Netthiem, a good friend of ours from Australia, who happened to be visiting Canada at this time, to be Rolf's godfather.

When Rolf was about two years old I took him to the Maple Leaf Gardens where I was working on an outside broadcast of a game of ice hockey. Well, the skipper took one look at Rolf and said, *"He is going to be our mascot for the year."* He hoisted him on his shoulders and skated around the arena to the enormous cheers of the huge crowd that had come to see the game. Rolf was presented with a blue and white jersey and a beanie. As I worked for the CBC in Toronto there were many deals with the ice hockey club, fixing any mechanical issues at their stadium. Today, Rolf is now married to Beverly, has four girls of his own: Alexa, Ellie, Holly and Tess. He lives in Sydney.

Jacquie and Rolf in Canada (1958).

## My daughters, Sasha (1963) and Belinda (1964)

A few years following, Jaquie and I welcomed two daughters who were born in Australia. Sasha was born in 1963 and Belinda in 1964. Sasha got married and had three girls: Ayla, Bronte and Cleo. Later in life, Sasha remarried Benno Fenger and they both live nearby in Nambour, Queensland. Sasha and Benno are mental health nurses, doing a lot of good work for their communities. Belinda and her partner have four boys: Philip, Terry, Jack and Rolf. Belinda lives in Greenmount, Queensland breeding dogs and horses. She also works at a kindergarten.

Rolf, Belinda, and Sasha with our family dog, Búho, which means "owl" in Spanish.

## *Receiver Tower*

One other major problem that was feared by other workers was the broadcast and receiver dishes, which were set on a high tower about 100 metres above ground level. No problem in summer but come winter, it was a constant problem. The dishes were to a broadcast that didn't need much adjustment but the receiver dish needed to swivel 360 degrees depending where the truck was broadcasting from. This was important to be working effectively on Saturday nights, the Toronto Maple Leafs ice hockey games, especially when defending their titles.

In the morning, when the members of the master control found the receiver dish could not swivel or tilt, an urgent call went out to the maintenance department to fix it, NOW. Later, when I became established as the guy who enjoyed it, I was called directly and would get dressed in my Arctic gear. Warm woolly overalls, wool scarf, fur hat and mittens, with a toolbox in a big, heavy leather bag. The base around the tower was well fitted with barbed wire to prevent anyone stupid enough to want to climb it but had a wire door with a large padlock (a small lock would be very hard to undo with mittens during a windy snow storm). Upon entering there was a cord for the first steps, in which I secured my toolbox only to haul it up from the top. The cord was balanced with a small weight on the top end which aided in the raising of the box. The stairs petered out about halfway up and the rest was a ladder. The broadcast dish was set below the platform and did not require much maintenance, while the receiving dish was about 1.25m in diameter with a hook in the centre for focusing the signal. Below that was the mechanically controlled dish, allowing it to swing through 360 degrees about 30 degrees from the horizontal. A giant cable was attached that contained everything required.

Now the problem was that during a snow storm, the snow would build up around the swivelling equipment. There was a heater to melt the snow, but the water from the melted snow would find its way around areas that the heater couldn't reach and would freeze up as ice. My job was to connect a portable heat gun and gradually melt it. Sometimes that would take over an hour if the wind was blowing hard with falling snow. The top platform was approximately 2 metres square with a secure fence on the perimeter of the dish securing the area.

One frightening day I got to the top and found that the dish was frozen with ice and locked to the safety rail of the fence.

I couldn't reach the area so I climbed over the safety fence on the outside and crept along to the problem. I was wearing earphones and had a microphone on my lapel, being in direct contact with the lower operator in the master control room. As I chipped away at the ice, slowly getting most of it away from the safety fence, I advised the master control to move the dish in a clockwise turn away from where I was standing. Imagine my surprise when the dish started to turn anticlockwise! I screamed *"STOP,"* to the operator. The operator's face would have dropped when I nearly lost my grip on the fence... Had it gone any further, this could surely have happened. I struggled back over the fence to lay down for a while, recovering from my ordeal. I told the operator I was ok and he could move the dish to where it needed to go, finally making my way down the tower again.

Another problem that occurred during my time there was that the power cable would freeze to the platform and when the operator started to swing the dish around it would put a strain on the cable, causing a disconnection. The dish would stop and although a signal from the receiver was working, the power to operate the dish wasn't.

On arriving up there with my toolkit I would disengage the plug from the end of the cable and remove the cover over the plug. Requesting the power to be turned off and I then had no electricity to heat my soldering iron. I had a small blow torch in the kit. To solder a wire to the back of the pin in a highly windy situation, and to keep a blow torch on heat, was quite difficult. When two or three cables had become detached it was difficult to figure out which pin to solder the wire to, especially when I had a choice of at least 25 wires. I would test with a volt meter and with a chart, hoping for the best. On the final test, my master control room operator would justify my anxiety. When I left the CBC that job ended too. They have a new tower in Toronto that is much higher but I believe is fully automatic, being able to operate in all weather conditions.

### *The Green Room*

Working at the CBC I enjoyed some interesting perks, for instance the Green Room. It was lavishly decorated complete with a bar, comfortable seating, an attendant and a kitchen to produce snacks. Its purpose was for any invited people to the studio to be screened or interviewed before they went before the cameras. They may have been singers, comedy artists or pianists, for example. Staff were invited to make these people feel welcome, offering them a drink, a snack and cigarettes to boot. When I found myself not terribly busy I would often be requested to front up to the Green Room to meet these new and hopeful artists. It was a lot of fun. I had the opportunity to meet some amazing people.

When musician Ravi Shankar (1920-2020) arrived in Canada he visited CBC. A good friend of mine, Bob Walker, an English actor, invited Ravi and his wife to my home for dinner. Ravi brought along his sister and they played Indian music together. He asked me to join in if I could play the drums with him, so I played on the bongo drums. Ravi went on to become world-famous, influencing many musicians, especially the Beatles.

Such powerful actors that excite me with their incredible appearances are Humphrey Bogart (1899-1957), Garry Cooper (1901-1961), Charles King (1895-1957) and John Wayne (1907-1979). For comedy I enjoy Charlie Chaplin (1889-1977), The Marx Bros (Chico, Harpo, Groucho, Gunmo and Zeppo), Bob Hope (1903-2003) and Bing Cosby (1903-1977). I loved; Bette Davies (1908-1989), Margaret Ruthaford (1892-1972), Barbara Stanwyck (1907-1990), Fred MacMurry (1908-1991) and even Gene Kelly (1912-1996). Paul Robertson, Peter Lorre (1904-

1964), Sydney Greenstreet (1879-1954), Boris Karloff (1887-1969) and Errol Flynn (1909-1959).

I met the actor Errol Flynn at CBC when I was working in the maintenance section, which was next door to the makeup room. Errol asked the makeup girl, *"Who was the Australian man next door?"* being an Australian himself. The makeup girl dragged me in to meet him. We enjoyed talking about Tasmania, especially Richmond, and chatted about my family's connection to the Richmond Bridge. He was most interested and we soon became friends. Sadly, in 1959, he died shortly after I had met him. During my work at CBC, I also had the opportunity to meet Elizabeth Taylor (1932-2011) and Richard Burton (1925-1984) as well.

## *Learning colour projection in New York*

At CBC during the late 1950s, television broadcasts were still in black and white, with colour being a frontier yet to be fully explored or even developed. CBC sent three of us to learn about colour projection from the Recording Company of America. One of the other fellows was also Australian, from an outside broadcasting (OB) division, mainly broadcasting sports. I went to the second-highest floor of the Empire State Building in New York, to their studio. Two or three people were happily dangling their legs over the edge of the building, overlooking Park Avenue. I believe it was about a week's worth of learning where I gathered a lot of information about colour projection. Lighting was the most important thing, and this was my field of knowledge.

I had a friend at the United Nations so during my free time that week he gave me a personal tour through the building to his office, overlooking the Hudson River, giving me a complete run down on what each person was doing at the UN. We enjoyed a coffee together in the canteen.

### *Meeting Prince Phillip*

At CBC I met Prince Phillip (1921-2021) when I was asked to show him the new colour projection camera. He seemed quite interested in it and we both enjoyed our meeting. Later I set up a school for technicians to learn how these jobs were really simple and fun to do. One of the technicians opted to come with me, as I taught him how to maintain the equipment. He enjoyed working with me but soon became bored when I was sought after by a firm to manufacture electrical goods.

While working at CBC I was headhunted to work in a manufacturing plant. A fellow from Dominion Electric said, *"We've heard about you and the things you can do. Our factory is at Kitchen and Waterloo; before the war, it was called Berlin."* The twin towns are on either side of a mountain in the province of Ontario. He asked me what salary I was on. I said, *"Thanks but I'm quite happy here."* He was offering me a production manager position with a substantial salary increase. I spoke to Jacquie about the offer. She had different ideas, wanting more money to rent a nice house and raise Rolf, so she asked me to accept the offer. I was persuaded to agree with her and phoned the company to accept.

# Dominion Electric Company

This company had several production lines. They manufactured refrigerators, record players and what have you, and I was to get my own office. My job was to watch out for when the red light started flashing, move swiftly to the production line and find out why the production stopped. I had to sort out the problem and fix it as quickly as possible. I stuck out this new job for only about three months until the managing director had different ideas for me. As we were chatting one day, he said, *"We live in a rather exclusive suburb. There is a nice house for sale, just across the road from me. It will be nice and handy for you."* I thought, no harm in looking. He said, *"While you are at it... why are you driving an old Dodge car? if you get a new car, you can park in the executive car parking lot."* He reminded me further that every Saturday, he had a little function at his house. *"You and your lovely wife and child can come over and spend the afternoon there if you like."* I relayed all his messages to Jacquie and she said ok.

I felt I was being coerced into doing something I was advised to do, and knew I was a born leader, who disliked being led by others. I distinctly remember it was this subject that I had my first argument with Jacquie arguing where we should live. We had three choices: we follow the recommendation to move near my boss or move on completely to either New Zealand or Australia. Eventually, I was able to compromise with Jacquie by going to New Zealand to see her family, on the way back to Sydney, Australia.

By now, we had enjoyed living in Canada for three years. I contacted my boss and told him we were planning to move back to Australia and made no excuses for not liking the job or anything, I only gave him family reasons, that we were going back home. Within a week or so we were flying back to Sydney. We stopped in at Los Angeles, Hawaii and then in Fiji, our third stop, where I got arrested.

On the plane, we wrote various letters and cards to people we were going to see when we arrived home. At the airport, I thought I could post these letters from Fiji. We were told that we couldn't leave the airport, but I didn't stop to think and walked across the street. A policeman stopped me and asked, *"Where are you going?"* My reply was, *"To the post office, to post these letters."* He asked, *"Do you have a visa to come out of the airport?"* I told him no. *"You are coming with me to the police station,"* he said and proceeded to charge me with penetrating Fiji without a visa. *"My plane leaves in 30 minutes,"* I exclaimed, to which he said, *"That's your problem, not mine."* Anyhow, after a lot of talking to and fro with the airport officials, I was finally escorted back to the airport to catch the plane. Jacquie asked, *"Where the bloody hell have you been?"* I simply laughed, as the plane was due to leave in five minutes or so.

We flew from Fiji to Auckland and spent three weeks with Jacquie's parents. It was here that Jacquie's father, Jack Crum, took me down to the brickworks where he worked, offering me a job there. I declined the job offer and Jacquie was very unimpressed. She certainly had a temper with me from time to time and I didn't enjoy that side of her. Most other times we had a marvellous relationship, enjoying each other's company immensely.

# CHAPTER 7

# Australia (1961)

Upon our arrival in Sydney, I assured Jacquie not to worry, as I had the accommodation all sorted. We arrived at the shed and Jacquie took one look at it and said, *"You think I'm going to live in this shit hole? You are mistaken my boy... you can put me into a decent house."* I replied, *"What about just one night, it's a lovely cosy little place?"* Jacquie's stern reply, *"Not for one night, it's a shit hole, get me out of here!"* So, I went directly to an agent and took the first house that was available to rent at Bellevue Hill.

I had a job all lined up... so I believed. My step-father knew the general manager of the Australian Broadcasting Commission (ABC), Sir Charles Moses. He introduced me to Charles, promising me a good job. Charles was known for bringing overseas Symphony Orchestras to Australian television. He became a big name in the industry, particularly during the 1950s-1960s.

I went up to the ABC headquarters, in Elizabeth Street, Sydney, to locate Charles. Upon meeting him I thoughtfully explained what I had been working on in both England and Canada. He replied, *"Very interesting. Go and see the head of the studios in Artarmon, North Sydney, and they will find an occupation for you."* I arrived at the studio and happily discussed the various parts of the studio and equipment that I had worked on. However, I was advised that although they would have no trouble finding me a job, I may have to wait some time to get a role that met my ability. They explained the company already had a few of 'my people', two camera men and a master controller. Television, by then, was up and running, so I was offered a cable puller job, which consisted of guiding the cable behind the camera or microphone. This was way below my dignity and in my mind, I said, *"You can stuff that job up your jumper,"* and politely declined the offer. My job search continued and it didn't take long to find a very interesting job.

To this day I have kept most of my career references. They give me great pleasure, reflecting upon my wonderful working life experiences. I am proud to say that I even have a reference from the Premier of Tasmania.

# Gollin & Company, the vario-klischograph

My cousin, Bill Angove, had a friend working for Gollin & Company. They were getting new high-tech equipment from Germany. I was hired to service these new machines. The Sydney Morning Herald bought one before they had arrived in the country. The vario-klischograph machine was invented by German engineer Rudolf Hell (1902-2002). It made printing blocks electronically, and I helped introduce the vario-klischograph into the Australian press industry.

A coloured photograph would be laid facing down on it and at the other end were three plates: yellow, red and blue. The machine could spread the ink from the three coloured plates, at the same time and a colour picture would be the result. Before that happened, colour photography would have to break a picture down into three basic colours. This entailed making separate plates, laboriously making one plate at a time, then raising each plate with acid, cleaning that off before making another plate and finally ready for printing. This new technology changed the printing world.

This vario-klisograph caused the closure of many shops around the cities that had made plates for the printing industry, for example those that printed brochures, newspapers or books. All you now needed was one machine.

When we unpacked the first vario-klisograph we soon figured it out and got it working. My colleague, Bob Crew, wondered *"What shall we print?"* I suggested Bill Angove's picture and because he knew Bill, he happily agreed. We put it into the machine ready for printing, in one colour only. It was a trial run and I have the very piece of material from that moment in my collection.

After a year or so, however, it was discovered that one of the managers had been fiddling the till at Gollin & Company, so we were all soon laid off. While this was being settled, I had to look for another job and tried Channel 10. The director asked me an unusual question, *"What frequency do we broadcast on?"* This completely stumped me and because I didn't know the answer, he believed I shouldn't work there. Then I found out much later, Mega Hertz at Channel 10 is 209.25 Vision Carrier MHz, 214.75 MHz Sound Carrier, but it was a bit late.

# Sydney University (1961)

## *Mills Cross Radio Telescope development*

It was in 1961 when I saw an ad in the paper *'Technician for the Astronomy Department, offered by Professor Harry Messel'*. The ad wanted a technician to mechanically design and supervise the construction team for the Mills' Cross Radio Telescope. I got the job and enjoyed it immensely, being on the staff in the Department of Physics at the University of Sydney. I learnt so many interesting things and met some very interesting people.

### *The Badham Building*

The physics building was originally called the 'Badham Building' and was constructed to the specifications of Sir Richard Threlfall (1861-1932) in 1888. By 1925 a second floor was installed to house the expanding departments. In 1940 the Professor of Architecture, Leslie Wilkinson (1882-1973) decided to alter the building further to create a Mediterranean style of architecture. I had an office on the second floor at the western end of the building. My office was next to the office of Professor Dr Charles Watson Munro (1915-1991) who had been the Director of the Nuclear Foundation at Lucas Heights.

I felt privileged to be working alongside incredibly interesting people. Quite often in the late afternoon, Charles would have tea and if I was still working, he would invite me to join him, relating many fascinating stories about physics and nuclear activity.

I was one of the engineers who worked on the then world's largest radio telescope of its time, the Mills Cross Telescope. The two arms of the telescope were 1500 feet long each, that formed a cross shape, to produce a fan beam in the sky. Professor Harry Messel (1922-2015) was the head of that department and wanted to build the radio telescope. A colleague, Arthur Watkinson, worked with me on this project. I made the parts for the telescope.

### Speed of Light with Professor Warcup

One of the most interesting times I remember had to do with the speed of light. Professor John Warcup (1918-2013) wanted to use the new physics building, which was finally finished sometime later with the addition of a second tower. He had calculated by measuring the interior hallways of the building that, including the tower, it came to exactly one mile. His idea was to place a mirror above and below the stairways at each end of the building and if he started the light source on the platform, at the top of the tower, it would be possible to measure the speed of light.

A very accurate clock was purchased: a naval clock set in a case with the usual gimbal for use at sea. However, it was modified by placing a very small mirror to the second hand drive shaft. A microscope fitted on the side of the clock was focused on the mirror. The setup was placed on a platform in the tower.

The light beam moved down the hallway to the next mirror, which was at right angles and passed the beam down to the next floor. It continued in this manner before returning the beam back to the mirror in the tower, next to the clock, where a reading could be made by watching the second hand. You can imagine the shock and horror, as eyes were focused on the second hand of the clock to measure the time delay when they discovered both light and reflection occurred at exactly the same time.

Light travels over 300,000 kilometres (186,000 miles) per second. Too fast to be seen. So, they went about their business to do other things and the apparatus was dismantled, finishing up in Professor Munro's office. He said, *"Would you like the clock?*" I was more than happy to secure it and is still in my possession today with its little mirror from the experiment still attached (but the microscope stayed behind).

It was an honour to meet physicist Dr Wernher von Braun (1912-1977) who specialised in space engineering. He showed an interest in the workings of the Mills Cross Radio Telescope and I was happy to demonstrate. Wernher was involved in engineering bombs that were used during WWII. The American government hired him after the war to head up the NASA rocket propulsion program, which resulted in the first moon landing.

## *Why is it so?*

By the time we had completed the designing phase of the telescope, we had to go down to Captains Flat, Bungendore, NSW (near Canberra), to put it all together. We had so many students in the physics department to help us; it was wonderful. My role was to show them how to join the wire mesh to the main frame and supervise to ensure they were doing it correctly.

While working on the Mills Cross radio telescope it was regarded that some of the mechanical components required plastic parts. Not knowing the first thing about plastic, I sought out several books on the subject. Putting this knowledge into practice I was able to make dies of various castings and apply them to making what was required. The engineers in the physics department had an old portable manual plastic moulding machine that I overhauled and got working. Nearing the completion of the telescope, a number of small parts needed to be made by outside contractors. Quietly I bought a similar machine and quoted on some of the plastic parts, which I made at home – and made quite a small profit from the venture.

### *Murder Mystery (1963)*

On New Year's Eve 1962, Dr Gilbert Bogle (1924-1963), a Commonwealth Scientific and Industrial Research Organisation (CSIRO) nuclear scientist said, *"I know things they don't."*

Dr Bogle had a love affair with the wife of Geoffrey Chandler, who worked for CSIRO, having the same equivalent position as me. There was a party at Chatswood, hosted by a CSIRO photographer. Mrs Chandler told her husband she would be home late. Together with Dr Bogle, they went to Lane Cove

River, where they experienced vomiting with diarrhoea and were found on the river banks dead. Their bodies were analysed but there were no abnormalities found.

I had an office near Professor Charles Watson-Munro (1915-1991) who was running the Lucas Heights Nuclear Facility. Dr Bogle was soon to be going to the United States to work on the development of the atom half-life. Mrs Bogle believed her husband and Mrs Chandler were killed by the Russians, who didn't want the Americans to have this information. The agent suspected of killing them was purported to be highly lethal but leaving no trace. Around this time, a Russian in Canberra was deported along with his wife, who was dragged off a plane. They were both convicted of spying. To this day, politicians want to suppress theories that point to Russia at this sensitive time. The murder mystery was the talk of the town for many years.

# 1st Restoration: 120 Jersey Road, Woolahra (1962)

The first house I restored was 120 Jersey Road, Woollahra, Sydney in 1962. It was a terrace house that had been made liveable by the brothers of the men responsible for the rebuild after cyclone Tracy (1974) in Darwin. They decided to move to another suburb in Sydney and the house was very cheap. We had saved quite a bit and with a small mortgage, we moved in. We ripped out marble surrounds and cast iron fireplaces to be fully restored and replaced into their original position. Walls were repainted and spaces enlarged to house one new bathroom and one new kitchen, shared by all. We converted many of the rooms into rentable rooms, setting up a thriving boarding house. We advertised to executives to be near work.

Firstly, Ian moved into the single bedroom, as he was now divorced from his first wife, Bea. Ian was my best-man at our wedding in London. Bea later became godmother to Sasha, and still to this day, I see her. She was a celebrated ballet dancer in the USA before returning to Australia. She was working at Scotts College when Rolf went to school.

The second bedroom we rented to a Spanish family; Julio, Maria and their six month old daughter were our first renters. We had noticed quite a few Spanish migrants had freshly arrived from their homeland after the problems they'd experienced during the Spanish Civil War. Julio was a chef at the newly formed Spanish Club at 227 Liverpool Street, Sydney. The club had a restaurant, dance floor and bar. Julio was elected to be the head chef and Maria waited on tables. When they set up the club they wanted members, so Jacquie and I joined, becoming officially the first and second members of the Spanish Club, which we maintained for a number of years. Julio certainly was a good cook and we loved his Spanish meals.

Our next tenant was a jet pilot who worked for the Royal Australian Navy, situated at Jervis Bay. Although he rented the room by the week, he only used it on the weekends, spending a lot of time bringing home very nice young ladies, but some years later he married. After retiring from the Navy, he became a commercial pilot for Qantas.

Bill Angove introduced us to an advertising executive, Bill Parnell, who took a room in our boarding house with his girlfriend. Bill was also a double bass player and played every Friday, Saturday and Sunday evenings at a hotel in Kings Cross. On Sunday mornings he volunteered his skills and played guitar at Ted Knoffs' Chapel (now The Wayside Chapel) to lighten the atmosphere. The chapel was a safe community place for people to break their drug or alcohol addictions. I was dragged along at first but soon became a very keen counsellor there.

About a year and a half later I found Maria on the doorstep of our house with her head in her hands crying gently. When I inquired if there was anything I could do, she said, *"Yes, maybe you can."* It appeared that when they left Spain to come to Australia, her parents were to follow and they patiently waited but the dictator of Spain, General Franco (1892-1975), had suspended her father's passport for some political misdemeanour. Maria said I could perhaps write to General Franco requesting a pardon for her father to come to Australia if I could guarantee he would never return to Spain. Well, this I was prepared to do, but knowing full well that it would likely be a complete waste of time. Nevertheless, I did it. Several months later a letter came from the Spanish Immigration Department asking me to sign a declaration that I would abide by my letter. This was completed and shortly after, Maria's parents arrived in Australia. They were so happy and all lived together in the one room. Big Mama, as she was soon to be known, took over the kitchen and provided some of my most memorable wonderful Spanish meals. The parents would get up early and I would drive them down to the markets for fresh food. It wasn't long before they found their own flat in Bondi Junction and moved there. We were very sorry to see them go but within a short time, after raising Sasha and Belinda, we too moved with our own family of five and bought a house in Vaucluse, Sydney.

### *Designing Jewellery*

It was around this time, still deeply involved with the Mills Cross radio telescope, that I was looking for something to do at the weekends. One day, while sitting in my converted garage (as my workshop was growing slowly), a scrap box was lying on the floor. This triggered my idea to start making shapes. Laying out the pieces on the bench, I would conjure up what I thought to be smart necklaces for young ladies to wear. Just a little offering of second-hand teak, say 30 x 60mm, with a piece of polished brass or copper and later silver, of any particular shape, a hook and chain – and voila, a necklace. At this time Jacquie had now befriended Germaine Rosenblum, daughter of Myer E. Rosenblum (1907-2002) a notable lawyer and Australian sportsman and when she saw these first models, I had made she inquired how was I going to sell them. I don't believe I had even thought about that yet. She asked, *"Can I sell them for you, as I know a lot of little outlet shops and many of my friends would love one too?"*

The polished stone would look good against the wood. The petrified wood had been brought down by the Gadigal people, the traditional custodians of what we now know as Sydney, to the beach to open shellfish and the like. It was evident by the huge number of middens (indigenous natural resources used for hunting, gathering and food processing, i.e. shells or bones) that the Aboriginals had accumulated many natural resources. (Way into the future I was to own land at Cloudy Bay on Bruny Island, Tasmania, where I had piles of middens on my property. I protected them as I was the caretaker of a very precious piece of Australian history.) However, being ignorant at this earlier time in Kangaroo Valley, and not knowing too much of the history, we used rakes to search for chips of petrified wood. How wonderful they looked, all the colours of natural wood, after being calcified. Back at my workshop I had a lot of fun, reshaping and polishing the gems. I was doing a great trade with my jewellery work.

In a magazine I had started reading about the tin industry in NSW and a place called Tingha, where quarries of quartz could be found in an old mining bed. I decided to go on a road trip with Ian to explore the area. On our way there, we stopped at a river that was known for having diamonds in the mud that were apparently thrown back into the river. On the river bank we discovered many old broken sorting tanks and remnants of camp sites. We made camp using the best-looking position and went to investigate with our panning dishes but as night fell nothing was found. A storm appeared to be coming up so when the first raindrop appeared I decided to sprint to the car, whilst Ian stayed by the fire. Minutes later the lightning struck a piece of metal nearby and suddenly I had Ian sitting beside me with a cup of coffee in his hands. We both figured it must have been one leap to the car, one leap through the window, and he was beside me. I don't think he had ever moved so fast. We camped the rest of the night in our car.

The following day we arrived at the mining sands, where tin had been mined in the past. All abandoned now, it was just about 4.5km out of Tingha. After a short exploration we found quite a few large pieces of smoky quartz in the sand of the old river bank, enough to fill two swag bags. A successful trip!

During the next day of fishing, some foraging and dredging in the middle of the river I experienced a Willy Willy (Aboriginal name for a whirlwind or dust storm). I explained it to Ian as we stood there, enveloped in the dust storm as my sunhat sailed high off into the sky. Boy did he get a good laugh out of that.

On the way home, while crossing the river, we got stuck in a sand bank and were unable to go either back or forward. As we set out along the road in the hope of finding someone living around here by the time it was dark, and whoopee, we saw a light in the distance. The house was not more than a shed, with an old chap and his dog. We never found out what he did for a living but he was eager to help when we explained our problem. He slung a coil of rope across his shoulders and we set off, wondering how this was going to work We thought he'd got another vehicle near the sandy ridge but when we got there, he tied it to the front of the car and with a bit of help from the engine, he pulled us by hand back onto the road. He would not accept any money or even a lift back home. I will always be grateful for his kind gesture.

We got home safely with our bags of smoky quartz. I had such a lot that I shared it among the other enthusiasts at the lapidary club, and some of the finished items were fantastic. I cut a few and mounted them on pieces of sheet silver and also pieces of wood like cedar, poplar and red gum. I thought these were quite beautiful and soon ventured into vitreous enamel on copper as decorative pieces on timber.

When I was experienced with the enamel I put about ten pieces in a bar and onto a high shelf in a box, and completely forgot about them. Some time later it was time to move on to Vaucluse for a whole new life, leaving behind the box.

Moving on, about 58 years later, I was in an op shop in Nambour, Queensland, browsing along the shelves when I came across a very familiar box. Well, it was familiar, but I had no idea as to why. I opened it and before me were the 16 pieces of enamel jewellery. I was so taken aback, wow, and why, *"What are they doing here?"* The shop assistant had no clue as to who would have donated them but she insisted I take them as a gift, so I went but not without making a suitable cash donation to their charity.

The only memory I have, apart from putting these ones into a box, was that I had done the same thing with other pieces of jewellery I had made but somewhere I still had them, although I haven't seen any of these pieces since.

# Laurence & Allumask

Upon completion of the telescope project, I was lured away from the university by one of the companies, Laurence & Allumask, that I had had dealings with. They invited me to become their factory manager. We were making die casting with zinc and I could see a lot of the items could be improved if they were changed to plastic. I approached the management to press my ideas but I was formally told *"We are die-caster people"'* and that was it. I did this for about a year but things were often going wrong and I was blamed for all the problems. I soon discovered that Mr Laurence, the general manager, was coming onto the factory floor and changing the production systems, unbeknown to me, so we lost production very quickly. I had an argument with him when I found out. He was interested in taking on my job and reassigning the production of certain lines. It was my role to explain the loss of production, in particular to the general manager of Holden for the Holden parts we made. I happily dobbed Mr Laurence in and promptly quit my job. A customer from Laurence & Allumask had heard about this and asked me to join his firm to make parts in plastic, however, I decided to set myself up in my own plastics business.

# CHAPTER 8

# TEB Plastic Enterprises (1965-1973)

I bought my own injection moulding machines. I made die castings, including toolmaking plastic products converting many products into plastic materials. For the name of my company, initially, I simply wanted to use 'Plastic Enterprises' but that name was already established. Then I was considering THB or TAB in front of Plastic Enterprises. Here I ran into a little trouble, as I was approached by an existing TAB (Totalisator Agency Board) a betting agency in Australia. They threatened me, but didn't offer any money for me to not use the word 'TAB'. Accordingly, that is why I changed the name to 'TEB'. I had one employee, but by the third year, I had two.

TEB Plastic Enterprises quickly picked up a lot of work. This was all performed initially in my garage in Vaucluse, not an ideal place for a plastic company so I found a small place in Redfern. I bought a new big plastic machine and I was on the way, soon, to be teaming up with another company nearby. This company was run by a Dutch Jewish gentleman who had trouble communicating in English. He had suffered the holocaust and met his wife in a concentration camp. They went on to live in New Israel but after hearing about Australia, they immigrated here. The couple started a small moulding business making toilet roll holders and were keen to expand their business. At this time, I came across another fellow from the rag trade and he was looking for plastic shirt buttons. Well with my new friend who I called 'Solly', we designed and made dies for all kinds of buttons and made millions of them.

Apart from the buttons we made a great variety of products, mainly for industrial purposes but there was one exception. I had been told of a Belgian product that was being sold in Europe. It was a plastic container with a packet of ground coffee which could be placed on a cup and filled with boiling water. This produced a superb cup of coffee. On seeing one of these contraptions, I redesigned it so that it could be used over and over again, simply by refilling it with fresh coffee. I had it working well and started to sell it widely, especially to

Nock and Kirby, the retail store. They employed a spruiker who called himself 'Joe the Gadget Man'. He had a regular weekly program on TV and would demonstrate tools and kitchenware so people could see what was on sale in the store. It wasn't long before he discovered my coffee cups and would start every show making a cup of coffee while demonstrating his wares. However, after a couple of months he called me in to say he wanted to be paid – handsomely, I might add – for demonstrating my cup. I refused and he started to run down my product saying it was too expensive to buy, and that it was rubbish. This affected the sales and I ultimately had to drop the product.

I was expanding my business, taking up another floor in the building I was now renting. I employed an apprentice, a toolmaker, a secretary and five women doing assembly work. I had several partners in business with me.

# 2nd Restoration: 122 Hopetoun Avenue, Vaucluse

We sold our terrace house and I sold my only painting by Sir William Dobell to buy a house in Hopetoun Avenue, Vaucluse, overlooking Parsley Bay with glimpses of Sydney Harbour. There was not a lot of renovating but I did tackle the underneath of the house, which was built on a steep slope, exposing quite a bit of room under the verandah and into the next space that was near the laundry. Rolf eventually took over the new room as he had been sharing with his sisters. There was a timber garage extending out from the house in the front yard. This is where I started my plastic business, in there with a small lathe and a second-hand moulding machine. There was plenty of storage for raw materials and an area for finished goods but it was pretty crowded.

### *Aminta and Julia, the pilots*

Our neighbour was the well-known Judge Mant. He rented out some of his rooms to a young lady in her late teens who had just arrived from Canada. Her name was Aminta Hennessy – yes, a member of the Hennessy Brandy Company. When I got to know her, she came over one evening asking about aeroplane wings because she was studying to become a pilot. I said, *"That's a funny thing, I happen to have a book on that topic,"* which I was able to lend to her. Aminta was absolutely delighted to have this information that helped her to pass the exams to become a pilot.

Aminta received her pilot's licence at Bankstown airport but was not satisfied as she wanted to fly commercially. In those days, TAA, Ansett and Qantas didn't employ women pilots. She searched around Australia to find a company that would accept her so she could get a commercial pilot's licence. Aminta finally found that Connellon Airways (later Conair) would accept her if she passed the commercial pilot exams. The airline operated from Darwin to Adelaide and out to the bush towns. She worked as a barmaid in Alice Springs until she had enough money to pay for the exam. She sat for the exam and passed, becoming the first woman to fly with a commercial pilot's licence in Australia.

When Aminta vacated her flat in Vaucluse, her friend from England, Julia Clifton Brown, moved in and we befriended her also. Julia wanted to get her pilot's licence too, so she sat for the exam and passed. Being able to fly solo, she said to me, *"Right you're it."* I was her first passenger. We flew out of

Bankstown airport, flying all over Sydney; it was a big thrill for me.

Julia became good friends with Jacquie. They stayed friends until Jacquie died. Sasha was a little girl in the earlier days living in Vaucluse but did develop a close relationship later in her life with Julia. Much to my delight their closeness continues to this day. Julia flew for various companies doing surveys and more, and later married another pilot, Bryan Laver.

They retired and moved onto a settlement just outside of Newcastle, near Morpeth. The settlement owns the land but provides an earthy communal type of living arrangement, growing their own food and supporting each other in various tasks or needs. I was up there in 2023, for Julia's 80th birthday. All the locals came and contributed food, and a well-known musician brought along his band to play music for us. The musician came back to Julia's house, staying on for a few extra days and playing enjoyable music for everyone. In Morpeth we had a good time exploring the rustic old township. Shops were selling old wares; it was a very interesting place. Both Sasha and I continue visiting Julia but sadly Bryan passed away in 2023.

We soon discovered that we also had a famous artist living two doors down; the wonderful Pixie O'Harris (1903-1991) who painted the children's wards in many of the Sydney hospitals. She also wrote many children's books.

### *Erica and Jane*

I became close to my two nieces, Erica and Jane, when my sister-in-law, Grete (short for Marguerite), was left when her husband walked out. Erica was crushed in a car accident and not expected to live but she pulled through her incredible ordeal. When she was recovering from her accident in hospital I lent her my lucky stone. I still have it in my bedside drawer. She went on to have her own baby and ran a marathon.

### *Harold Holt, USA Ambassador BBQ (1966)*

US President Lyndon B Johnson (1963-1969, 36th USA President,) came to Sydney. It was Saturday the 22nd October, 1966, when Jacquie, Rolf, Sasha, Belinda and I stood on the roadside of the new Anzac Parade to get a grand view of the president as he passed by in the motorcade, travelling along the streets of Sydney.

Later the same day, President Johnson flew to Canberra for an informal BBQ at Lanyon Station (now known as The Lanyon Homestead at Tharwa) hosted by our Prime Minster, Harold Holt, as well as Edward Clark, the Ambassador of USA and his wife. Coincidentally, my mother was the housekeeper of this homestead. She was asked to organise the BBQ. At the end of the evening the president came over and congratulated her on a very pleasant afternoon. Some time later she received a letter of commendation and some photographs that had been taken at the event. She swiftly ripped up the photographs in exasperation because President Johnson had recently sent rifles for the IRA (Irish Republic Army) to shoot the southern Irish rebels in the rebellion of her homeland.

## *Divorce*

By 1971 Jacquie and I went our separate ways and got a divorce. Looking back, when my father divorced my mother in 1945, he gave her everything he owned, including the entire contents of their rented house, and paid alimony to her. A similar situation occurred to me upon the breakup of my marriage, but the worst thing was walking away from my children. Jacquie later moved to Queensland taking the children with her, who were now teenagers. It was the worst moment of my life and I felt suicidal.

At this time, I was running a company, having a number of employees, and felt obligated to them to keep going. I set up a camp at work, literally, by screening an area off where I could cook, eat and sleep. There were big financial outgoings to my former wife. It was a disappointing and very difficult time.

### *Sydney Opera House (1973)*

I made some items for the new Opera House in Sydney that was being built in the 1970s. They were plastic rollers in nylon which enabled the large glass windows to be put into place safely. My contribution to the building gave me an invitation to the exclusive Grand Opening Party. The Opera House was officially opened on the 20th October, 1973.

### *Poker Machines*

During the 1970s Nutt and Muddle were makers of coin operated poker machines. I did a lot of work on the balancing wheel adjusting the cog holes, which would produce a win-or-lose effect for the player. Casinos could choose the outcome of the players by manipulating the desired end result.

## The Old Stage Coach Hotel

After some years at the home of TEB Plastic Enterprises in Elizabeth Street, Redfern, the Lebanese church population decided to turn an empty block of land into the site for their new church building. Upon excavating the land, it was found there were the remains of an old hotel. On doing a bit of research myself, it appeared that a stage coach had a departure hotel there. Bulldozers were removing the top soil and soon exposed a wine cellar. The builder filled the back of his car with the old bottles, called it a day, closed up the cellar and went home following his discovery. After I had finished my day, I had dinner and prepared to inspect the cellar myself. I rang Jacqui to come over as she was still living in Sydney at this time, knowing she had a keen interest in old bottles. As it turned out, she was extremely interested; she kept the whole find for herself. We discovered that being underground, the corks were rotten even though there was still wine in them (but the wine was way off-drinkable). Between the two of us we collected some 100 bottles. Jacquie simply got in her car and drove home. The bottles were never to be seen again by me. I do believe she sold them promptly. The very next day the dozer driver ploughed the remains of the cellar, destroying everything including its vaulted ceilings. Now all gone and smashed into tiny pieces.

### *Factory Fire*

During the following weeks the excavators continued to dig down beside our adjoining wall. It was a few weekends later I had a visit from my mother to see how I was going. It was a Sunday, so we enjoyed having our lunch together at my factory. I took her down to Central Station to catch the train back home to Faulconbridge in the Blue Mountains. When I returned to the factory and opened up the front doors I was met with the sound of fireworks. *"Damn,"* I thought to myself. Some kids must have got in and were exploding crackers. But no, the sound was coming from the lift shaft. Trying the lift to the first floor, it wouldn't work, so I took to the stairs. By the time I got there the lights went out and I could see flames from inside the lift wall. The handy fire extinguisher beside the lift entrance made very little difference. I moved to my office with just enough light to find my torch I grabbed the phone and dialled 000 to summon the fire brigade, then quickly got the hell out of there. It was about 6.30pm. The fire was mainly in the roof and the top floor. The firemen arrived and called for two machines with long extension ladders. By midnight the worst was over. I drove around to the house of my ex-business partner, Solly, and begged him to give me a bed for the night. He was only too happy to oblige but his wife found it very difficult to be civil to me. The next night I booked into a hotel in Redfern Street and that was my home for the next two weeks with no business income.

Mr Varga, the factory's landlord, raised stink demanding I continue paying rent. I thought this was completely unreasonable but hoped his insurance would pay for all of the expenses. I moved back in when the power had been restored but found a huge mess from the water and fire soot that had left marks on everything. I had to replace a lot of moulding powder and the stock had been severely contaminated by the water. About a week later I was able to turn on my machines to resume work again. Over the next few weeks Mr Varga's insurance took a massive number of photos and completed a detailed list of all the damages. Unfortunately, Mr Varga insisted that the builder next door was responsible and should foot the bill, not only for me but all the other tenants in the building who suffered losses. We as a group appointed a solicitor to secure our claims, making sure we were all well compensated for our trouble. Sadly, the solicitor tried to contact the builder only to find he was unregistered. He had left the state and was unable to be located.

Just to annoy me further I lost a good part of my stamp collection in this fire. It was nicely assembled in a box on the floor. It was in Taree, a few years later where I tried to recommence my stamp collecting.

Mr Varga appeared in a court hearing to determine who was going to be responsible. The jury found him to be innocent and laid it in the hands of the solicitor. Next bit of good news, the solicitor took me and the other tenants for a fees only involved in finding the cause. A new builder came on the site and put up a reinforced wall next to the adjoining building and built a very fine church on the site. Two of the tenants in our building were bankrupted. The others managed to get by as I did but the little nest egg I was saving for a new machine did not eventuate.

### *The Broken Promise (1973)*

One of our jobs included using polycarbonate sheets to make replacement glass in all the gauges used by NSW Government railways in their steam, electric and diesel engines. This was in case of an accident, so the driver's cabin would not have glass flying everywhere. By 1973 I was approached by the Australian Government to think about stabilising a country town, by relocating my business, to prevent young people from leaving and going to the city. I was promised full assistance in relocation funding on building new premises, free phone calls and substantial payments to rehouse myself. What a great idea, especially as I was becoming very dissatisfied with one of my partners and his son, who I was sure was having a personal dip into my button stock supplies, selling them off as his and pocketing the sales.

So, I decided yes, I would move, and after cruising around NSW I chose Taree as my next town. As part of the Government Agreement with the Department of Decentralisation, created by the then Prime Minister, Gough Whitlam, I was to work for a year before I received any compensation from the decentralization program. That was ok; I borrowed sufficient funds from my bank, who thought it was a grand plan. The local council was very helpful, giving me a superb block of land on which to build a new factory, and they even let me have an architect to design the whole thing. In the meantime, they supplied me with an old building previously owned by the county council electricity department, as they had just moved into a new premises.

One of the nice things that happened in Taree was when I was asked by the local magistrate if I would consider employing a young man that was serving time for robbery as part of his sentence. On this I fully agreed, and the well-mannered, trustworthy man was a splendid worker; he only needed to be given a chance. His stealing was out of pure desperation and he regretted it terribly.

When a year went by I had a call from the bank to say I needed to get the government to repay the debt. I'd fully set up and was enjoying the benefits of decentralising my business. But it was a Friday when I heard from the Department of Decentralisation, and was informed that the plan had been shelved because it was not central to the Albury-Wodonga industrial area. They asked if I would consider another move. Without haste I rang my solicitor and that afternoon sat in his office. The Government Agreement was only a handshake and was worthless. I had no paper work with the promises of the funding, and therefore my factory park was foreclosed by the bank. The following day, two semi-trailers came and carted off my entire factory. I was completely broke, with no company and no compensation. I returned to Sydney a very sad and disillusioned man.

# CHAPTER 9

# New Beginnings (1975)

### Marian

Ian Hamilton's second wife, Sue, introduced me to this lovely woman, Marian. We hit it off immediately and were soon planning our wedding. We were married on the 7th May, 1975 and lived in Cook Road, Centennial Park, near the Sydney Showgrounds. We lived there for only about six months before we bought a very run-down terrace house in Redfern.

Our wedding day, Marian and me (1975).

Our wedding reception (1975).

## 3rd Restoration: 215 Cleveland Street, Redfern

The Sydney Mail Exchange had used two terraces in Redfern as their offices before their new building was built. The terraces were full of old wheelbarrows filled with dried cement, sand and what have you. Thus, the terraces were very cheap to buy, considering how good they actually were having been hand-built during the 1880s. There were original cedar doors and floors throughout.

Since my setback in Taree I had found a job as a sales representative for Edward H O'Brien, the producer of the Yellow Pages phone book. It paid good wages but entailed some arduous country trips. However, I found time to roll up my sleeves and completely rebuild the house to get it into a shipshape condition but it was a big challenge.

When standing in the living room one could see planes fly over through the many holes in the roof. I installed a new roof, windows and staircase. I also removed the deteriorated wall plaster to expose interesting brick work in the kitchen, laundry and the outside toilet. (There was another toilet in the upstairs bathroom.) We turned the front room upstairs into our bedroom.

One of the decisions I made was to buy a large collection of cedar floor boards from a church in Glebe but I decided that I didn't need them, so I laid them down in the wine cellar, kitchen and laundry in parquetry patterns. The parquetry floor was sanded and polished to perfection. It looked very good and I was extremely proud of my job.

At the front of the house, I used temporary timber posts to support the front verandah. The original pillars were made from cast iron columns with Ancient Greek designs but I was unable to find suitable replacements. So, I travelled down to Wagga to a cast iron foundry and commissioned a pair to be cast and fettled. The pillars were transported by train to Darling Harbour. I picked them up from the railway station, which was at the beginning phase of being entirely demolished. With much grumbling and with some neighbourly help, we got the pillars into position. They are still there today.

Before leaving Sydney in 1997 this house was being sold and I went to look at it, to see what changes had been made twelve years later. The agent was most enthusiastic about the kitchen floor, which was supposedly laid by the current owner. I did take my photo album with me, showing him how I had laid the floor myself and that the owner was exaggerating a bit much. The agent was surprised, hugging me and congratulating me on an excellent job.

### Dick Smith and Bob Pennicott

I had recently been commissioned by a new company to sell plastic machinery. Having just got married to Marian, and still being on our honeymoon, I was asked to see a customer while there for the plastic company. Arriving at the front door, I found the customer on his hands and knees, and he asked *"Do you know what I'm doing?"* It was Dick Smith, the entrepreneur and aviator, playing with a train set on the floor. We later shook hands and I commenced my successful business sale. Many years later I caught up with Dick in Adventure Bay on Bruny Island, with Bob Pennicott, owner of the Wilderness Journey company. Dick was known for flying solo in a helicopter, circumnavigating from east to west around the world in 1995. Bob had cruised his 4.5 metre inflatable dinghy (boat) around Australia in 2011 to raise awareness and funds for eradicating polio. Bob is now operating the Bruny Island Cruises and Seafood Restaurant.

## Saving a Life

It was around 1975 when Marian and I were staying at her parents' home, in Kingsford, while they were overseas and we were getting a new roof on the Cleveland Street terrace. One day there was a loud banging on the front door that had me running quickly to see what was happening. An Indonesian girl was holding her baby, screaming, "*She's dead!*" Wondering what I could do, I grabbed the baby, who was looking quite blue, and started to breathe air into her mouth. We made our way to the living room and I directed Marian to call an ambulance while I offered CPR to the limp, tiny body. I alternated breaths and cardiac pumps, and within minutes a little colour appeared back in her face. I checked for any pulse rate but continued with my CPR procedure. Marian called out that delays would prevent the ambulance from coming within 20 minutes. Thankfully, the baby was slowly returning to pink but there was little movement in her hands. I instructed Marian to drive to the Crown Street Hospital ASAP with the family, while I stayed by the phone and rang the hospital to be ready to receive them and then cancelled the ambulance. At the time, it seemed the fastest way to get the baby to the hospital, while I phoned ahead. Off they went with the baby, now coughing and crying in the car, very much alive.

About an hour or so later they all returned happily with their baby declared as fully recovered. I never did find out what caused the baby to suffocate in the first place. The mother was all smiles and she gave me the grandest warm hug. That evening her husband called in to also express his feelings of immense gratitude. He asked if there was anything he could do to repay us. I felt that nothing was required; we were just so glad that we were home and able to help with the emergency. He explained that he worked for a car yard, inquiring if I would be interested in a new car. *"Hell yes, but money is tight,"* I declared. He informed me a practically new car had just arrived back in the yard as it had been returned by a café owner who decided the vehicle was too big for their needs. It was a Toyota Crown and I could have it at a highly reduced rate as it was now considered a second-hand car. Both he and his boss were so pleased with the outcome of the day's emergency they made a very special offer to me. Marian and I happily accepted. The day turned out so well and was definitely worth celebrating with a glass of red.

## *Howard Silvers' Chandelier (1977)*

Howard Silvers was a firm selling electrical items, wholesale. Howard himself asked me to make him a fancy modern chandelier for his dining room at his house in Darling Point. I charged a lot for this job, $800, and installed it in 1977.

## *The Fireman's Helmet (1978)*

One night Marian got up about 2am and noticed smoke bellowing up into the sky not far from Redfern. She quickly woke me saying there was a fire in town. I got out of bed and said we should go and have a look at it, picking up my camera. We jumped into my car and drove five minutes to Ultimo, where the fire occupied an entire city block. Thankfully, the fire was in an old abandoned building that use to store wool, and the 300 residents who lived nearby were already evacuated. We went around three sides of the fire and then came across the fire brigades, who were trying their best to put the fire out, although it was too far gone now. We watched for a while, took photographs, and finally decided to just go home. We never thought more about it and both went to work later that morning.

I had the pictures developed and found them to be remarkably interesting, as it was rare to capture such a big blaze along with the hardworking firemen busy putting the fire out, so I took them in to show my work mates. A woman said to me, *"Wow, this was a really big fire and I know all about it. Do you see that car outside the building? It was my father's and he is the superintendent of the Fire Department for NSW. He told me that although it was a remarkable fire nobody was there to take any photographs of it. Would you be willing to share these with the Fire Department?"*

I was only too pleased to share them and soon received a thank-you letter. Later I met the chief superintendent of NSW Fire. He asked if there was anything that I would like in return for my donation. Quickly thinking, I said, *"I've always wanted a brass helmet,"* hoping they might be able to dig up a couple for me. In those days the brass helmets were no longer used.

It wasn't long before I got a phone call from the chief. He exclaimed, *"This is most unfortunate, all the fire helmets that we had in the stations were collected, then destroyed. We've had many requests from governments around the world for these brass helmets too. We've gone back to the original manufacturer, a company called 'Bell'. They still have all the equipment to manufacture them. They are making only four duplicates, but we have asked for an extra one."*

Nearly six months went by and I was contacted to go down to be officially presented with my fireman's helmet. I was only too happy, and the firemen were smartly dressed in their uniforms at the presentation. I was thanked profusely for the photographs as they had been sent around the world. I took the helmet home and placed it on my stand. I was very proud. Soon after, a magazine came out from the fire department with all the pictures and the story behind them, with a note thanking me for the photographic efforts. A Canadian fire insurance company even bought some photographs to use them in their marketing.

As time moved along I kept looking at the fire helmet. Many friends considered me very lucky to have it. However, I was a little disappointed as the helmet had never seen any fire action. Amongst the many bits of correspondence regarding it, I received a letter from the Museum du Fire in Switzerland asking if it would be possible for me to find in a second-hand shop a genuine brass helmet, as they were not on the list to get a reproduced one. The moulds had been destroyed after the last batch of reproductions were made. So I went out shopping, searching through all the Sydney antique shops, until one day I found one. It was priced at $1,000 and I bought it immediately. Getting home I sat the worn, grubby helmet next to my shiny, newly reproduced helmet and thought, *"They are going to get the new one and I will keep the old one."*

The museum in Switzerland now displays the reproduced helmet that hasn't seen a fire or any action. You couldn't tell the difference between the two, only by their visual condition. I believe the helmet I have is an officer's but I don't know the rankings of firemen. My helmet has two or three dots on the back, indicating the rank. I still have copies of the fire department's magazine showcasing the photographs I took that early morning. I also still have the original fire helmet the museum paid me to locate and purchase for them.

Fire News Magazine front cover, my photographs (1978).

Fire News Magazine back cover, my photographs (1978).

# Palingenesis Antique Restorations

*"Terry Butcher of Sydney, a metal craftsman whose restoration skills include almost anything from a thimble to a mangle".*

> METAL CRAFTSMAN – PUBLICATION
> THE AUSTRALIAN OLD HOUSE CATALOGUE, 1984, THE COMPLETE 'WHERE TO GET IT' GUIDE FOR THE HOME RESTORER BY IAN EVANS AND METHUEN HAYNES. (PAGE 72).

By the late 1970s I had established a very profitable business dealing antiques and restoring furniture. I named my new enterprise, 'Paligenesis' meaning to re-create or birth something new.

## 4th Restoration: 56 Pitt Street, Redfern

In 1978 Marian and I were walking down Pitt Street when we saw a 'For Sale' sign in the front yard of a terrace house with a broken window. The yard was rundown and overgrown, with a rusty iron fence. The original owner had become so old, he was unable to care for the home fully, so he put it on the market. He had lived there for 60 years.

We immediately contacted the agent at Bridges Real Estate and arranged a deposit of $10,000. The house price was $90,000. Before accepting, he told us another buyer was offering a $11,000 deposit. He asked us to match it or lose it, so we matched it and the house became ours. The agent had gazumped us, lying about having two buyers. Today this house is worth around three million dollars.

Francis James Gilmore was the Mayor of South Sydney in 1900 and had lived in the house. It was opposite the Redfern Town Hall. At this time the electricity commission was set up in Sydney with a power house that used steam in a coal-burning building just off Redfern Street. The project was to provide electricity for lighting at the Redfern Railway Station. Being a mayor, Francis directed that his house be wired for him also to use the surplus power from the new technology. Therefore, this became the first private house in Sydney to use electricity.

The house included a stable and ballroom. I was able to use this extra space to work on my antique furniture restorations. But when we moved into the house, it was in such a mess. The previous owner and his wife had made the living area in the ballroom and were eating out, not using the kitchen. Most of the floors had been destroyed by white ants and the skirting boards hid moisture in the walls. There was no running water and the bathroom was as inhospitable as the kitchen. The tessellated tiles on the laundry floor were all broken up.

So just as they had lived in the ballroom so did we to start with. We set up our bed and wardrobe with room for our furry friends, the cats and dogs. We used a kerosine stove to do our cooking. I had the water running the next day.

We bought a gas stove cooker to replace the huge 8 burner that filled the entire kitchen. It was so ancient I decided to store it, telling Marian I would repair it when I could get spare parts. Yet this became very difficult, and the old stove sat in the stable until we left Sydney, sadly never to be restored. In the kitchen there was also an original wood-burning fuel stove. This I did manage to repair and get working after many years. Finally, there was an all-brass, gas-burning, water heater over the sink. I got this fixed but had to replace the mantle. I chose a second-hand cedar mantle, one from a junk yard, which looked great after I had redressed it with French polish.

The kitchen required repainting so I carefully removed the dark cream paint and found, to my delight, the original green paint with a blue line pattern at the wall's central height. With the help of my friend, we made a stencil and repainted the kitchen as it would have originally looked back in 1897. We bought an old pine table and stood it in the middle of the room with a huge pine, glass-fronted dresser along the wall adjacent to the stove. Marble tiles surrounded the old wood stove and kitchen sink. At last, the kitchen was functional. We just had to replace the floor in our bedroom, then we were able to move out of the ballroom.

We loved our Victorian breakfast room; there was not much that needed doing there. The two doorways from the room had coloured glass inlays. One led into the kitchen through a small hallway and walk-in pantry and the other led out to the courtyard.

One night we were awakened by a loud crackling and after racing downstairs we found the whole of the breakfast room ceiling had collapsed, leaving behind a terrible mess. The ceiling had been water damaged from the bathroom, which was directly above. A new ceiling was fitted with new wiring to the chandelier, and I found a magnificent ceiling rose at a small firm that was still making them. Later all the rooms were refitted with a variety of Victorian roses. I had a huge job later in restoring the bathroom floors, relaying them with the original marble tiles. I pulled all the marble tiles off the walls, cleaned them up and carefully replaced them back into their original position.

Over time I had to replace the floor boards, including the framing and floor supports. There was an air space from the wooden floor to the underside of the flooring of about 75cm. The skirtings were removed and silicone was pumped into the walls to remove the moisture and seal it for the future. It was a very costly operation. These skirtings were all cedar with moulded tops. They had all been polished and finished with lacquer, so they needed to be sanded back again, cleaned and re-polished. This was a very arduous job but the results made it all worthwhile.

*Sydney newspaper (circa 1995) – publication,
A Redfern renovation; Living in the '90s (1890s) written
by Geraldine O'Brien*

"Terry Butcher seems to have mastered the difficult art of living in an authentic period house - a skill few get the chance to achieve, and fewer still would choose to practice long term. It means, after all, no dishwasher and, in the Butchers' case, a cast-iron bath placed publicly if picturesquely under the datura in the garden. "Quite private, quite private", Butcher insists. They (workers in the large commercial building next door) don't start till after seven). The bathroom - pure turn-of-the-century, like the kitchen - is still being renovated: the marble wall tiles having been polished and replaced, the elaborately decorated toilet awaiting reconnection and the bath still needing some work.

But for inveterate collectors and hoarders such as the Butchers, progress is not linear but a labyrinth of byways and detours - to collect some genuine old Holland blinds, say, perfect for the upstairs drawing room; or to acquire the job-lot of moulding planes which sit on a tray-mobile in the breakfast room; to find the kitchen's corner rack for the copper saucepans or the collection of flat-irons in the ironing room above the wash-house.

This is a happy story of the right house - an 1896 terrace in Pitt Street, Redfern - meeting the right owners, people who, in Mr Butcher's own words, "didn't want to live in a National Trust property but wanted this house. "When we saw all the wonderful things here - well, we just had to keep them and adapt them to our way of living". The house was built in 1896, one of three. Because the builder was to occupy it, he "went crazy on it", says Butcher, making it wider than the other two, almost double their length and more elaborate.

At the back is what must be Redfern's only surviving private ballroom, semi-derelict (the renovations haven't progressed that far) but still with its delicate painted swags and garlands, visible on walls and ceiling. "When we got the house, there was practically nothing done", Butcher says. Indeed, the previous owners had progressively retreated before the onslaught of time and pressing repair needs, gradually abandoning the main body of the house to the white ants and holing themselves up in the ballroom. This policy played havoc with the floorboards, but it did mean the house was spared numerous layers of "fashionable" paints or generations of "fashionable" renovations. Not only the ballroom decorations survived: original fireplaces also remained, as did most of the doors. The walk-in pantry between kitchen and breakfast room still. Had its mesh panels. In the upstairs salon, the floor was still japanned. In the scullery, white ants had chewed away the years, causing the collapse of the tiled floors. Terry Butcher sat down and picked out every remaining terracotta or cream tile, which have now been reassembled in situ. In the kitchen, which had been repainted, he scraped back to reveal the original colours. These have been reproduced, leaving a small square of the original paintwork behind a dresser. A steep flight of wooden stairs, with the treads worn in the middle, leads from the maid's room to the scullery below. "There's been a lot of traffic up and down those stairs in the last 100 years", Butcher points out.

These are the visible legacies of the past, but Butcher says, there is also a hidden heritage, equally eloquent testimony to the story of the house. "At some time during the 1920s the house was rewired and they didn't rip out the old wiring, they just put in the new and disconnected the gas. I rewired into the original places". And in the hall, he has carefully restored the cedar-and-marble wall mounting which proudly displayed all the switches and fuses. Like the indoor *toilet and the single power point in the upstairs salon, it bespeaks a house which was, at the turn of the century, the last word in fashion and technology. And although there are now copper pipes in the house and a Miele washing machine in the laundry (along with the copper and original tubs), this is essentially the house of 1896 with its interior functioning pretty much as it did then. Its survival is a small Sydney miracle. With luck, any new owners (after the Butchers move to Tasmania later this year) will find the same comfortable familiarity with the old, the faded and the authentic as their precursors".*

I continued with my antique restoration business using the stable until we moved to Bruny Island, for my retirement. We lived in this restored terrace for about twenty years.

From this location I enjoyed meeting many high profile politicians, of the day; Paul Keating (Prime Minister), Neville Wran (NSW Premier), William George Hayden (Governor General) Bob Hawke (Prime Minister) and his wife, Hazel and Warren Anderson (businessman), just to name a few.

# Black Tie 50th Party

Marian turned on a wonderful Black Tie Birthday Party for me at our home in Pitt Street with the following guests:

- My best friend, Ian Hamilton and his wife, Bernadette.
- Grete (Marian's sister), Paul, Erica and Jane made my birthday cake with fifty 50c pieces, adding a razor and mug for decoration on top.
- Professor Stevens from the Engineering Department of Perth University.
- Moira Pointer and Bill, her partner, the Jockey.
- Stuart Symonds (he had the biggest collection of Pianos which were donated to Perth University) and his partner, Cliff.
- David Rossell and Jeanie Kelso, both opera singers from the Australian Opera Company then working at the Sydney Opera House.
- David Kinsella, (Kinsella Funerals) a well-known organist (and harpsichord player) at Sydney Town Hall. He could also play a harpsichord (a much earlier predecessor to the piano). I met him through Bill Bradshaw.
- My mother, Rita, of course.

- Owen (chief pilot for TAA was later killed in an air show) and Frieda O'Mally.
- Grahame Dunne, my dentist and his wife, old friends of Marian's

David Kinsella played the old piano while we had Jean and David singing a couple of songs from 'The Merry Widow' an operetta. My musical friends often entertained us at Pitt Street. I will never forget that, it was delightful.

## *Door Knobs and Chandeliers at Admiralty House (1983)*

When Princess Diana and Prince (now King) Charles came to Australia in 1983, they stayed at Admiralty House, Sydney. In anticipation of their visit, I was employed by the government to make new door knobs throughout the entire house, as I was a sought-after antique dealer and restoration manufacturer in Sydney. I cast them into brass then polished them until they glistened before fitting them into all the doors. Secondly, I was to clean and check the chandeliers hung in the guest bedroom, to be used by Prince Charles and Princess Diana. I sat on their bed to admire the bedroom's freshly cleaned and polished chandeliers.

The house curator told me that the grand staircase and the doors would be painted white. I almost fainted and told a good friend, Stewart Symonds at the National Trust, of her decision. The curator was quickly replaced and the grand staircase remained in its beautiful red cedar French polished state. I made the grand chandelier that hung over the staircase. I was not allowed to take any photos around the house but I do have photos of the door knobs being polished at home.

## *Hyde Park Barracks Clock (1984)*

This public clock was commissioned by Governor Macquarie and designed by colonial architect Francis Greenaway in 1816. It was important for organising the convict workforce as only wealthy people could afford a pocket watch in the early colonial period. I had the privilege of working on this clock, having to stop it for the first time since it was installed in 1819, so I could repair it. My name is stamped on the bearings to show the authenticity of the repairer. This is Australia's oldest public clock that is in full operation today. The building was fully restored in 2020.

## *Lantern for Bob Hawke*

Bob Hawke was the 23rd Prime Minister of Australia, from 1983 to 1991. I was asked to make a lantern light to hang outside his door, at the back of Kirribilli House, the Prime Minister's official residence, next door to Admiralty House. Upon completion, the job of fitting the light was given to an electrician for installation. Bob Hawke loved it but when his wife, Hazel, came out to view it she said, it was all wrong and had to be removed. I did get well paid but do wonder what happened to it in the end.

## Old skills can still be hired – publication 'Your Home', a bi-annual publication by Moira Maguire article.

"Any old iron - it all comes the same to a master craftsman like Terry Butcher, who is a dab hand at working with metal. A fine example of Terry Butcher's craftsmanship, these scales, which date from 1870, were in pieces when they were brought to him for restoration".

## *The Redfern Floods (1988)*

One of the most extraordinary events I experienced was in the heart of Redfern, on the corner of Elizabeth and Redfern Streets. It was a Sunday morning at about 5.50am, on the 27th July 1988, when a water mains pipe burst. Water poured down Elizabeth Street towards Waterloo. The police came in their numbers and prevented any traffic chaos, cordoning off all the streets around the area as the water levels quickly grew. Members of the council arrived to see what they could do.

I was planning to take a shower but found no water available and then heard sirens. I decided to quickly get dressed. As the fire brigades were going past I wanted to follow them with my camera. The commotion was not far from our house, about two blocks away. I made my way down to find a small crowd of firemen, police and council conferring. They were trying to speak to someone from the water board to come quickly and turn off the mains. Shortly after two members of the water board arrived but didn't know where to turn the water off. TV journalists started to arrive with their gum boots on. The water became so deep that one couldn't walk down the road. As I took photographs of the flooding and watched all the goings on, a rubber duckie boat came past with council workers. They travelled down the road to where the houses being flooded were.

I walked through the adjoining Redfern Park and down to Waterloo where water had entered the houses of several streets. There were two government ministers who were emphasising to those affected by the flood that they would be well compensated for the damages. A discussion with the water mains engineer disclosed that the underground plans for the water mains were in the office of one of the senior staff members, but they couldn't locate it. It was now nearly 5.00pm. The water had moved across Redfern Park. Much of the water was flowing into the drains but they couldn't keep up. By 6.30pm the valve was finally located and the water flow was turned off. Now work started to replace the broken section. The following morning we had running water again. It took over six months to get the locals back into their houses with brand new furniture, fridges, etc.

### *The Legend Café (1991)*

I vividly remember this moment in time, it was the 6th April 1991 at the Legend Café in Sydney. I was enjoying a meal with

Bill Bradshaw, Paul Keating (Australian Prime Minister 1991-1996) and others. To mark the occasion Paul began recording a video, he framed me directly and in response, I swiftly grabbed my camera to capture a photo of him filming me. Paul and Bill were good company and shared my interest in antique restorations.

Paul Keating filming me (1991).

# CHAPTER 10

# My Collections

*Tool's Paradise - Good Weekend - publication*
*The Sydney Morning Herald, Good Weekend,*
*15 September 1990.*

"*Collector, Terry Butcher doesn't believe in restricting his fossicking to a single category - say, French clocks, garden gnomes or Georgian silver. Instead, he is fascinated by the prosaic objects of 19th and 20th century living, the essential gadgets and tools and the myriad ways inventors designed them. At his home in Sydney's Redfern, where, between visits to junk shops, he makes a living repairing metal antiques. Butcher has box upon box of can-openers, scores of corkscrews, drawer after drawer of shaving implements and an impressive array of hand tools. His wife, Marian, with whom he shares a dilapidated, six-bedroom mansion, collects linen and lace. Beside the half-dozen bedrooms, the Butcher's house boasts a ballroom, stables and servant quarters - it's the perfect venue for an unusual collection.*"

Collecting for me started at an early age with little model cars. My parents went to a ball and on the table at the dance was the menu supported by a small model car tin. Probably made in Japan. When it was time to leave, my dad collected a number of these for me, which he displayed on the breakfast room table. I was overjoyed. I had a special place in the garden where I played with them so often that one day I asked my dad if he could get me another one. Over a period of time the three grew into a shoe box full, a dozen or more.

When the time came for Dad to take up his new job with the Perth Daily News, I took my first collection of toy cars with me and put some of them on the shelf above the hand basin in our cabin. Every bump of the ship rolled and so did the toy cars back and forth on the shelf. Dad insisted that while we slept they were to be put away. I was suffering from sea sickness for most of this journey and spent a lot of time in our cabin getting out my toy cars to play with.

The motive for collecting may start as an innocent hobby when finding something that is appealing to oneself. In 2005 ABC television program 'The Collectors' inspired many people to start collecting. It is believed one item is an object, two is a pair but three becomes a collection. With three pieces the collector may not feel satisfied and go on the hunt for more. At this stage it may still be regarded as a hobby, if the items are kept together where the items are either accessibly displayed or in an album for ease of viewing. This way the collection can be enjoyed by the collector or shared with others at any time. When the collection gets too large, it can be shared with the public, being placed on display for viewing individually or with other like-minded collectors. A public display may get recognition from a program like 'The Collectors' one day.

Some people are born collectors. I started off collecting pebbles and shells off the beach, and then I found coins, which led me to collect stamps. It was an uncontrollable experience spanning my entire lifetime, having developed this desire to keep things and not throw them away or pass them on. I enjoyed keeping them so I could always have them with me to look at any time.

# Meccano

I had to get rid of a Meccano set (a metal toy reusable construction set created in 1898) but if I could have kept it, I would still have it today. Meccano was the beginning of my interest in tools. It showed me the ability with my hands and mind to put a nut on a screw on a piece of metal and tighten it up. But I was so enthralled with my first collection, set number one, that I wanted set number two and pestered my dad until he bought it for me. For my birthday I got set number three and for Christmas I got set number four. I then asked my dad if I could have one box to put them all into. He asked a carpenter at the Perth Daily News to make a big tray with pockets all around to hold the screws, nuts and spanners. I was able to pick up the entire tray and put it under my bed, and I would get it out often to play for many enjoyable hours. I particularly loved making tractors and cranes with Meccano.

Over four years, during the war, I had begged, cajoled and pleaded, that I would sell my soul for the remaining sets of Mecanno up to set No.10. The No.10 set came in a big box with compartments of all the pieces, drawers for the screws, nuts and washers. At last I had a huge collection. It never left me but while overseas my stepfather sold it.

Playing with my Mecanno set in South Perth (1944).

Do your items simply sit on a shelf to be admired or do you say to yourself, "How long have these been around?" and timestamp them? You may categorise them into man-made or made by nature. To further your collection, you might like to display earlier models or other makes and pieces from different locations. If you have started to expand with this in mind, you can seek out more information by researching or joining a club that has already been formed for a collection of a particular genre. As your collection begins to grow your knowledge of the subject will increase, and so will the value, yes in a monetary way but also as a reference for you.

If you are confident to show others your collection, you will find no two collectors are the same. You can trade with others by selling or swapping duplicate and excess items. This will help you to grow your collection. Your collection, for whatever it is, is a piece of history and will no doubt add, no matter how small, to humanity's knowledge of the past. Almost anything can be a collectible item. Thinking first of man-made objects.

One day a collector may halt collecting, for three possible reasons: they may have lost interest, cannot find more items to add, or have decided to dispose of the collection. I am confident in saying that with the benefit of what you've already accumulated, your items will no doubt end up in someone else's collection. Their integrity will be preserved because you originally rescued them from oblivion.

Tens of thousands of tonnes are destroyed either by burying in landfill, recycling, burning or converting. I'm not suggesting for one minute that everything should be saved, no way, but I do hope that at least one of everything man-made is kept, and I'm sure that it will be. When I ran my own plastic products factory, over the years we produced thousands of different articles and in the end, I discovered I had only saved three or four. I know this story is repeated over and over again with the thousands of manufacturers throughout the world but no matter, I'm confident someone out there will have one of my plastic products, and another collector will have another one and so it goes on. As for nature, fossils, for example, will never be discarded once found. They will be passed along from collector to collector and the more interesting or rare fossils found will end up in public collections like museums.

So in my mind, all collectors combined have and will contribute to the preservation of wondrous artefacts. Like archaeology, where we scour the ground for evidence of a past life, treasuring every item found, even a small chip from a clay bowl, we can learn much from these tiny pieces of history. Future generations will not have to dig up the ground to see how we lived in this century, for all will be revealed from our collections.

My father would bring home an office diary for each year they were printed. I didn't keep a day-to-day report of what I was doing but stuffed the pages with bits of memorabilia about the life surrounding me. I did write down significant entries that occurred, for example when the Americans in Perth were being very generous, they gave us lots of chewing gum which differed from our local Wrigley's brand. The American brand was small flat ribbon gum wrapped in a colourful foil. I kept the wrappers, putting lots of them into my diary. Train, tram, and theatre tickets were placed in my diary also.

# Losing my early collections

When the war ended Mum gave me all the old ration books, which were used for food and clothing with lots of tabs still in them. I think there were some war-saving certifications in there too. They cost 15 shillings each and when the war was over we were paid much more. I believe they may have totalled approximately £4 in value after the war as the government were redeeming unused rations at the post office. I remember putting these old ration books into my box, next to my toolbox. I kept my big box of Meccano and left another box with all my Biggles' books, A. A. Milnes' Winnie the Pooh books and many other adventure books that I had collected.

On my return from Canada, all that remained was my toolbox, a dictionary and the Roget Thesaurus book. The rest my stepfather said was confined to history, as they were play things of my youth. He deemed I did not need them anymore so he sold them. I hated Peter from then on, and never got over the loss of my precious memorabilia and toy collections.

# My Collection List

Silver, old glass, books, Sydney Harbour Bridge memorabilia, old brass ware, antique furniture, old prints, old paintings, works of art, records, cylinder 78, 45, 33, CDs, tapes, elephant ebony and other materials, sporting items, darts, game boards, cards, chess, Ludo, scientific instruments, fob and wrist watches, mantel clocks, corkscrews, bottle openers, tin openers, shoe horns, nut crackers, knife sharpeners, pen knives, little openers, kitchenalia, scales, smoking requisites, small musical instrument, coins, rocks, minerals, semi-precious stones, fossils, bones, shaving equipment including razors and brushes etc., old cameras, newspaper cuttings, magazines, musical instruments, piano and organ, TOOLS; planes, screwdrivers, chisels and carving chisels, saws, saw sharpening, shoe making shoe repair and casts, leather work, farrier, rulers and tapes, vulcanising equipment, hacksaws, pluming tools including pipe threading, spanners and wrenches, tyre pumps, car jacks, valve grinding and tapping vices, post drills, levels, monkey and slipping gauges, callipers, hammers, hand drills, breast drills, bow drills, Cooper tools, axes and adzes, square and bevels, spoke shaves, routers, plumb bobs, measuring and metrology equipment, drafting and drawing instruments.

## *Deaccession Sale*

The Sydney Museum of Fine Arts and Sciences had a deaccession (to sell or dispose of items from a collection) sale. Marian was working for the museum, helping them sort out linens of which she was an expert when she saw the chest of drawers. She admired the goat's feet so much, and knowing it was going to be sold she wanted to buy it. Marian kept bidding for it and finally we got it for a good price, though a lot of money, and we were very happy with our purchase. This piece remained in our dining room in Tasmania until I moved to Queensland. I sold the drawers at auction, making a profit.

## Envelopes and Transcripts (circa 1800)

I have a collection of letters from Dr Burchell, my uncle. The envelopes have a seal on the back, which dates them 1820 to 1830. Postage stamps came out in 1840 during Queen Victoria's era, so these letters had none.

## Seeds from William Burchell (circa 1800)

My ancestor, William Burchell sent seeds he had collected on his travels throughout Africa in the early 1800s. I have these seeds in my collection and hope my family may take an interest in them.

# Sydney Harbour Bridge Collection

At Lavender Bay, on the lower north shore of Sydney Harbour, there was a minister who had a church positioned in an ideal location to watch the construction of the Sydney Harbour Bridge. The minister wrote about all the events that took place in Sydney. His sermons often changed topics with new sermon speeches relating to the general life stories about the Harbour Bridge construction. He even built a tower on the church's grounds to watch the building proceedings first-hand. They eventually allowed him to go onto the site, where he took many photographs of the construction process and took detailed notes to record the historical facts.

### *Sydney Harbour Bridge Spanner*

In my home was a large glass-fronted bookcase in which I displayed much of my bridge memorabilia. I had many teacups, saucers, plates, sweet dishes, some engraved glasses and about eighty souvenir teaspoons with enamel badges. I also had many souvenir puzzles and books on the bridge. In a wooden box housed a metre-long spanner that was used in building the bridge. In the bridge, rivet holes had a nut and bolt filled during erection, and large sections were held together in this way until removed and a permanent rivet was hammered into place.

Mitre long spanner used on building
the Sydney Harbour Bridge.

## *Sydney Harbour Bridge invitation*

About five years ago I rediscovered an invitation for Mr Herbert E Bellamy (1877-1947), an engineer, to the opening of the Sydney Harbour Bridge. I had acquired this piece of memorabilia in 1960, some thirty years after the bridge opening. At that time not many people were interested in collecting such items but I was delighted to have come across it. This is one of the ultimate pieces of bridge memorabilia. On the opening day, one would have to present the invitation to the authorities, and then be seated behind the premier of NSW. Mr Bellamy, having kept his personal invite, would have been very special. Many would have thrown the invitations away. I have since put this invite into a picture frame.

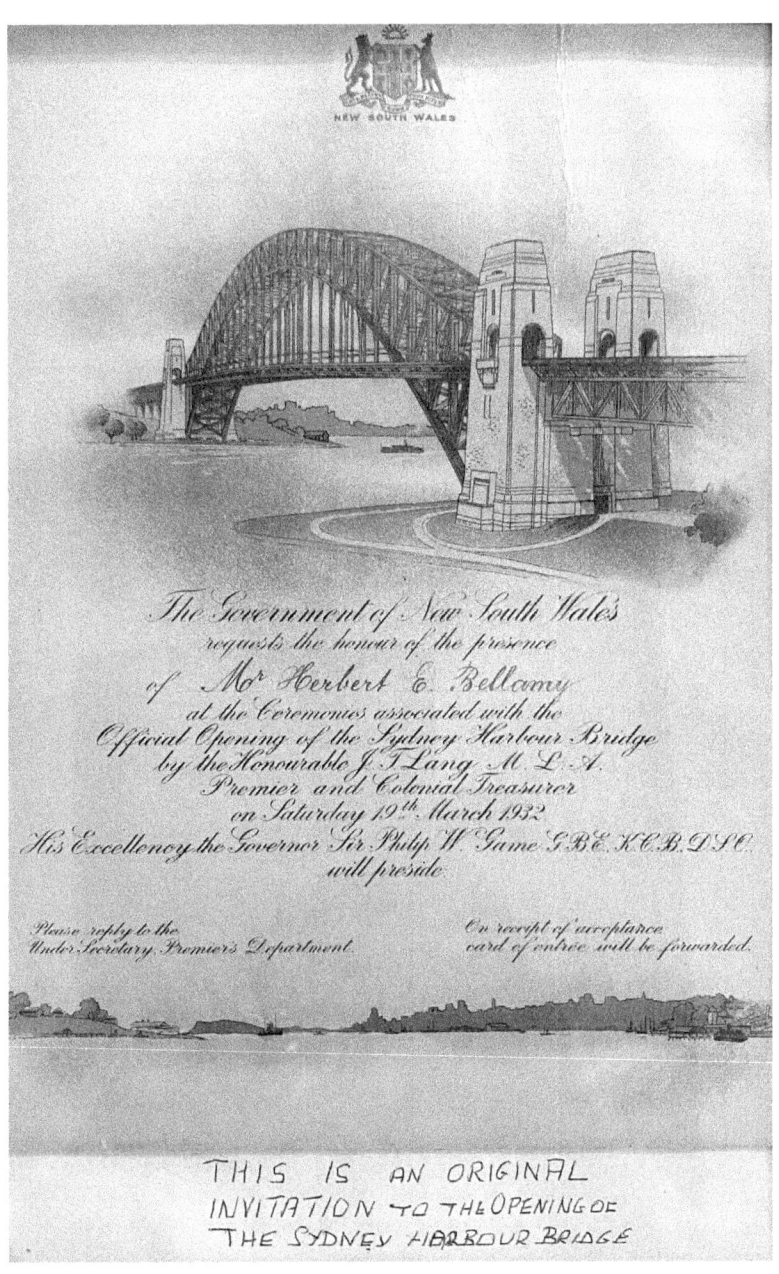

Sydney Harbour Bridge Opening Ceremony Invitation (1932).

I kept most of my Sydney Harbour Bridge collection in the dining room in Tasmania. All the walls were covered in many pictures: some photographs, some original paintings and two prints showing all the stages in the construction. For example an image of the opening (9th March 1932) by Premier Jack Long and the illegal opening by Captain Francis de Groote (1888-1969) of the home guard, who slashed the ribbon first. He was arrested and fined $5 for damaging government property. When I had my home in Tasmania painted, back in 1998, the painter I employed was Bob de Groote, a nephew of the now infamous captain.

### *Ceremonial Ribbon Cutting Scissors*

The original scissors used at the opening ceremony of the Sydney Harbour Bridge, in 1932, were solid gold with opals set into them. They are insured for approximately $1,000,000. Back in 1987, when we were approaching the 80th birthday of the bridge, the NSW government took a train around Australia showcasing all the memorabilia. The most outstanding feature was the gold scissors given to Governor Lang. They were manufactured by Angus and Coote, a Sydney jewellery company. They were made especially for the NSW government because they featured a design of the Sydney Harbour Bridge.

The government decided to go to a reproduction company, which required a very hot rubber mould to be made around the scissors, and someone became aware that this may destroy the colour of the opals. So they tried to figure out how they could make these reproductions. Previously to this, I had made some memorabilia items from the Sydney Harbour Bridge under similar circumstances, where I was able to make a cold mould, using artificial rubber that would not damage the opals. They asked for my help and I accepted the job to produce a pair of silver scissors, exactly the same, gold plating them and adding new opals. They would be placed in a showcase on the train. The original golden scissors were housed in a glass case, in the foyer of the NSW Government's building, Parliament House, in Sydney, where they remain today. I do believe the scissors were taken out of the case, and used only once, for the Premier to open the Harbour Bridge Tunnel. They have otherwise always been locked away, except while I had them.

On the day, I was waiting for the scissors to be delivered when a big armed guards' car pulled up outside my house. Two guards came into the house with this little box. They confirmed, *"You are going to make a copy of this?"* and I replied, *"Yes, that's right."* They proceeded to sit down, and I commented, *"You're not going to wait for them are you?"* The guard replied, *"We will wait until you have made the mould."* I advised them, *"That is going to be quite complicated because the mould takes a day and a half to cure, and you can't take it with you too soon. My suggestion to you is that I have a big German shepherd dog here, and he will have the mould in his kennel, where there is no chance of someone taking it away, and you can come back tomorrow afternoon to collect the scissors."* The guards made a few phone calls and finally agreed.

With haste I got stuck into the task at hand. In order to make a rubber mould it was necessary to cover half of the scissors in plasticine. I needed to make a rubber mould in two halves. I went up the road to the local newsagent, buying a packet of kiddies' plasticine in five different colours. I made a little wooden box and then made an impression of the square surface area with the plasticine, laying the scissors on top, pressing them deep, and then filling up the box with my rubber solution. It finally set hard. Taking it out carefully I turned it over to proceed making the other half of the mould. Upon completing the cold mould, I was able to take the scissors out, returning them to their original case and the armed guards in the afternoon of the following day. Upon their return the guards breathed a sigh of relief. It was fun to hold, clean and copy them. I have to mention this fact: when I removed the scissors from the mould, one of the opals came loose and I re-glued it in place. However, it has been my worry ever since, having not been able to find an original photograph, to check it was re-glued up the right way. Someone may pick up on it in the future, if they notice it is upside down.

Having produced a mould, I went back to the firm that was going to do the casting. They filled it with hot wax and when that cured, they put the wax component into a box of Plaster of Paris, sealing it all up. It was placed into a small furnace, melting the wax, and then the wax was poured out. The cavity was then filled with solid silver. When they opened the Plaster of Paris mould, there was a gleaming pair of silver scissors. We cleaned them up and then they were gold-plated. The new opals were duly fitted, and the scissors went directly into the display case. As far as I know, the National Trust still have possession of these scissors. Little did people know they were only a copy.

Just to make sure everything was perfect, I asked the castors to make two pairs of scissors, in case something went wrong. I kept the second pair of silver scissors and the mould. It was my intention at some later stage to have the scissors plated in gold and dressed up with some opals. However, I later thought better of that idea, not wanting to defraud the government, opening up myself to serious problems as to how I happened to have a duplicate set of scissors. I decided to leave them as they were when originally cast, carefully kept away. They will be a reminder to my family of the scissors and scissor mould samples when I have left this happy world. There were no contracts to sign from the government for my work. It was a normal thing to do at the time, doing a second piece, if something was very complicated or expensive.

# CHAPTER 11

# Collecting and Restoring Old Tools

### Tools are an extension of one's hand

You will find daily use of many tools, for example a knife and fork, razor blade or toothbrush. These are all tools; an extension of your hand that you can't do with bare fingers. Some take a fair bit of ingenuity to design, while others become just a fashionable item.

> METAL URGES – THE COLLECTORS – PUBLICATION
> CLASSIC AND COUNTRY DECORATING, 1993, WRITTEN BY
> ALEXANDRA NEUMAN

"What does a shopkeeper, an accountant, a blacksmith and a dentist all have in common? The unlikely answer in this case is that they are all members of a recently formed club for tool enthusiasts called the Trades Tool Group. Terry Butcher, co-founder and one of the *original 19 members, invited us into his unusual Redfern home to view the extensive collection of old tools that he has gathered over the past 20 years. Though Trades Tool Group is made up of collectors of tools from trades as diverse as shoemaking to dentistry, Terry's prime interest is tools used to fashion metal objects. He became so proficient In using them that during the 1970s he was well known in the antiques trade as the person to approach if a single chair castor, drawer pull or ancient lock needed restoring. "I think that the beauty of old objects is better appreciated if you know how they were put together and what tools were used in the process", explains Terry. He now not only collects old tools but restores them as well and if he is repairing a piece such as an old lamp or hinge, he always tries to do the job with tools from the same period. If a specific tool isn't available, he will often make it from scratch, using early tool manuals as a guide.*

*His workshop is hidden at the end of the garden and is, in his own words, 'ordered chaos' with a thousand one items – unrecognisable to the untrained eye – but a veritable treasure trove to those who know what they are looking at. And though he holds a physical interest in metalworking tools, you will spy at least six or seven woodworker's planes and the odd jewellery drill among the lathes and micrometers. All his extra-special pieces are kept in immaculate condition in a display cabinet in the dining room. Among his favourites is a late-18th century wooden brace, a Chinese marker that uses ink instead of chalk and a mid-19th-century square decorated with brass insets.*

*Collecting these tools is not simply a matter of possession, part of the fascination lies in knowing where and when they were made, exactly what they were used for, how long they were in production and what superseded them. The Trades Tool Group is invaluable for coming up with this type of information, with members being encouraged to bring unusual tools to meetings for identification. As one in the group was a former Australian tool manufacturer, a mystery find rarely remains a mystery for very long. Old books, manuals and catalogues are also useful for identification purposes and Terry has a large library packed full of such literature. With the world turning to computer-generated production processes, Terry Butcher desperately wants to preserve these vital pieces of our mechanical past and hopes the Trades Tool Group will be able to set up a permanent exhibition and workshop in the near future at the old Everleigh railway yard in Sydney."*

I didn't collect very much when living in England, possibly due to lack of money, it was a matter of survival. In Canada I worked on studio equipment. Here I had to buy new tools for that; pliers, screwdrivers and head torches, which form part of my tool working equipment. Those tools were different to my antique collection, which tools I don't use in work, generally. However, furniture rework, in antique, yes I did use some of the tools then.

When I came back to Australia I found time to browse around tip shops, junk shops, and the odd occasion at auctions, as the years went by I went to more auctions, they were the great supplier of antique tools over the years.

I have enjoyed collecting antique woodworking tools, which are now displayed in my home. In my childhood, my father had a lot of tools that he used, mainly for picture framing, and they intrigued me. I never owned such tools until I served my apprenticeship. The field of engineering I was in required a large number of tools, particularly measuring tools. There were also hammers and shaping tools. With each operation I was involved with I was allowed a new tool that was required.

A particular favourite tool was the metal stamp and hammer, for stamping numbers on a finished product. I was able to build up quite a collection by the time I finished my apprenticeship.

All of these tools I was collecting, for my personal use, which I could use every day, were all new and up to date.

We had an old Chinese fellow in our maintenance department. He was retiring and gave me a very old shoulder plane produced in the 1850s, made in Scotland or England. He said, *"You might find this useful. I've had it in my family for a long while."* This was the first time I had seen an antique tool and it was mine. This was a really big beginning of a lifelong desire to add to this collection and find out more and more about old tools.

This became a little bit of an obsession. Being in the restoration field I was able to get my hands on many more antique tools. I searched through second-hand shops or went to auctions in Sydney. I often went to the Tempe tip, the local garbage dump, finding tools people were simply throwing away. One day I had the luck of finding a large old wooden toolbox full of moulding planes. This added enormously to my collection because they were tools I didn't have more than one of. At that point I only had one moulding plane. Today I have more than a thousand.

I have book making tools and equipment from the 18th century to the 19th century. I know how they work; I can work them all.

I have a typewriter from 1900 when it was a new craft.

When I moved my business into Pitt Street in 1978, I was fortunate enough to have a spare room to put my old tools on display.

### *The Traditional Tools Group*

At that point I was fortunate enough to meet people from overseas who were tool collectors or tool sales people. Around 1990, I was talking it over with a friend of mine, Henry Black. We thought it would be a good idea to follow a group in Melbourne, that had just recently started a tool club. I was in favour of combining the Sydney and Melbourne clubs into one, but Henry said no, we can set up our own club. We set up a committee, electing him as president until we had our first public elections where people were voted in by members. It was on that occasion I got to be president of the tool club, called TTTG (The Traditional Tools Group) until I moved to Tasmania. We met a lot of people from overseas, sometimes at our invitation; other times they were looking for markets to sell their tools, mostly from France, England and the USA. I'm happy to say I'm still in contact with many of them today.

In Tasmania I built three sheds on my property. One to work in, one to put the surplus into and one was going to be a very big display shed. I continued going to auctions and op shops to increase my collection. Yet the display shed was not fully finished by the time I left, even after twenty years of opportunity.

### *Can Openers*

Back in the 1970s, when a friend of mine showed me a collection of very early and old can openers, I hadn't given it much thought about their history. We did some research, looking into why tins were made in the first place, well before the can opener came into business, and how they were invented. The early suggestion for opening a can was with a hammer and a chisel. An enterprising fellow was putting corned beef into cans and created a tool in the shape of a bull's head with a very strong cutting edge on one side. From that early beginning, hundreds of thousands of manufacturers from around the world have duplicated this innovation to produce even a very smart electric can opener. Today most cans are fitted with ring pulls.

### *Corkscrews and Bottle Openers*

I thought to myself corkscrews and bottle openers were of

similar value. I put two and two together, realising they are much the same thing, helping to open a bottle or a can but I didn't realise, when I started collecting them, that there is such a variety. I enjoy looking for new ones; I still keep finding shapes I haven't seen before or knew about, sometimes finding a similar copy of the ones I have collected. It has grown into an enormous collection. I find that other people are sharing my passion for collecting corkscrews and bottle openers. I desperately want a place to display them.

### *Shaving Gear*

When I was running my business 'Palingenesis' from Pitt Street, Redfern, I got a lot of visitors but there was one I remember because he was a bit different. I was collecting shaving gear and I gathered quite a lot from the Tempe tip shop. I had heard that there had been a period in the 1930s of a wind-up razor, shaped like an electric razor, but it never went on the market so I let it be known amongst my friends that if they heard of or saw one to let me know.

One afternoon I had a knock at the door. I opened it and was taken aback a little. There was a quite rough-looking character wearing biker clothes standing in front of me. Peering past him, looking at the road, sure enough, there was his Harley Davidson parked with all of its chopper dressings. He had heard that I was a collector of shaving gear and that I was interested in a wind-up razor. We introduced ourselves, his name was Max. I invited him into the house and suggested a cup of tea, which to my surprise he accepted. Going through the process of tea making, setting out cups and a plate of biscuits, he had by then taken off his goggles and helmet, along with his very large leather jacket adorned with chains.

His father had a wind-up razor and used it for many years but had since died. He didn't know what to do with it and asked if I would like it. Well, he did not have it with him and suggested I call into his place at Wooloomooloo. We agreed on a price and made a date for the following day. Wooloomooloo is a dreary suburb just behind Kings Cross and a little back from the Harbour. In his sitting room, he gave me a cup of tea in an enamel mug, but it was hot and good. He had shelves filled with all types of bric-a-brac. He had collected the items from friends or acquaintances and usually sold them on.

The razor was on this table on top of a cushion. As it had belonged to his dad he was loathe to sell it but insisted it go to a good home. He said he was happy with my home and myself, and therefore pleased I was going to take it. Over the years he visited quite a bit. Sometimes with something to sell but mainly just to enjoy a cup of tea together.

One day he surprised me by inviting Marian and myself to his wedding. His bride-to-be, Jacqueline, lived in a housing commission flat with her father in Surry Hills. The marriage was to take place in the common room of the flats. He wanted me as his best man. He explained to his friends that I was his closest, friend and he felt honoured by my presence. About a dozen people from the flats came and a local marriage celebrant performed the ritual.

After saying goodbye following the wedding, we left and never saw them ever again. I often wondered how they got on and where they went. He was a terribly nice man and I always enjoyed his company.

Max and Jacqueline's wedding day in Sydney.

# CHAPTER 12

# My Inventions

I did not get around to thinking of my inventions to discuss in this book until my daughter, Sasha, reminded me of a puzzle I had put together in the late 1960s. With a square and a piece of cardboard exactly the same size, the cardboard was cut into various shapes and laid out on the square. On rearranging all the pieces, it was always found to have one extra piece that never fit. I approached the John Sands Company, which manufactured and distributed board games in Australia from 1960 to 1990, as it was not the kind of thing I wanted to make. In their wisdom, unless I revealed how it was done, they were not interested, so I put it aside and continued to amuse my friends with it.

*Throughout my life I was only too pleased to solve other people's problems with my innovative designs. Of the various inventions that I created, some paid off and some did not.*

## Apprentice tool making

When I was a fitter and turner apprentice I was allowed to make my own tools rather than buy them. I still use some of those tools today. One is a modified tweezer for taking out metal splinters, which was an occupational hazard in metal working factories.

## CSR Development of Masonite & The Aerosol Canister

My friend's father, Len Lawrence, became an executive with CSR in the development of Masonite. They developed Tilux, which is Masonite scribed with grooves. Knowing my engineering skills, Len invited me to their factory to invent an automatic method for painting into the grooves. With much consideration, I was able to devise an aerosol canister that did this successfully. I was paid for my design but CSR patented my invention sometime in the late 1940s.

## Serrated Glad-wrap Cutter

Glad Wrap is now a well-known and useful product, but when

it first came on the market the firm selling it approached me to make a design for a suitable container to hold the film and access it readily, somehow cutting it into required lengths. At the time scissors were the only way to cut the plastic film. I made a suitable container to hold a roll of Glad Wrap and across the front I fixed a hacksaw blade. After unravelling the required amount, with a sideways motion it would neatly tear off. They were very pleased with my model and idea, paying me handsomely for making the design model. Sometime later the Glad Wrap Company packed the film into a box with a strip of serrated metal attached to the lid for ease of cutting the film. I did not benefit financially from the ongoing use of my innovation, as I never patented it.

## Lindemans Champagne Cork Remover

When corks were being slowly replaced from cork to plastic, I was approached by Lindemans Wines to see if I could make an opener for the new plastic corks for sparkling wines. They were the same shape as the old corks, including the wire restraining method to keep the cork in. So on with my thinking cap, I came up with the hard plastic opener that had an elongated hole to go over the top of the cork. It reduced the size to just under the head of the cork allowing the use merely by pulling upwards and removing the cork. *"Fabulous,"* said the board at Lindemans Wines, and they gave me a huge order to produce their lifters. However, before I could deliver their first batch they cancelled and went back to using cork, scrapping the idea of plastic. I noted a bit later a corkscrew manufacturer took my idea and produced one in pressed metal similar to my original design.

I was made an Honorary Member of the Corkscrew Club in the United Kingdom in the 1980s. I believe the gentleman from the English club was holding my letter moments before he passed away. My letter was telling him of my interest in corkscrews and my extensive collection. He was delighted to have known there was an interest in the subject in Australia.

## Car Cigarette Dispenser

As I was a fairly heavy smoker in my younger days, I contemplated a self-serving dispenser that sat up under the dash in your car so when you felt the need to smoke, you merely touched the base of a little box and a cigarette promptly dropped into your hand. No need to fiddle with a packet of cigarettes lost in your clothing somewhere. I had a quiet moment and the problem was solved. Great, I thought to myself, now I need an endorsement from car makers, who all posed to me that I should get the final approval from the National Roads and Motorists' Association (NRMA). Yes, what a great idea. The Governor and his lackeys expressed what a fantastic idea it was, simple to use and taking no one's eyes off the road to open the container for a cigarette. The bad news was, that the NRMA, along with government agencies, were just about to launch a plan to discourage smoking with a quit-smoking campaign. So thanking me, they said it was just too late. In hindsight, I should have gone to a cigarette manufacturer; my timing was wrong.

## Hand Surgical Instruments

I knew a hand surgeon doctor in Sydney. I was able to create and modify existing surgical instruments for him in stainless steel. Some were made from dental instruments and others were miniature that helped the surgeon to operate on tiny bones found in the hand.

# Nasal Continuous Positive Airway Pressure Mask

Professor Sullivan of The University of Sydney was studying the reasons why people slept but suffered poor sleep. The mother of my then-wife Marian was married to Jim Bruderer, her second husband. Marian's biological father had died shortly after she was born. Jim was an engineer working for the university under Dr Sullivan at the Royal Prince Alfred Hospital. On behalf of Professor Sullivan, Jim asked me to have a think about a mask that covered the nose, in which airflow could be controlled.

I went out and bought a variety of masks: snorkelling, dust, painter, chemical and gas worker masks. I played around with all of them but considered the painter's mask and made several modifications to it. It had eye pieces, which I removed. The mask was still held in place with a couple of rubber bands to the back of the head. The mouthpiece let air in for breathing. It was now light to wear but would remain in place while the patient slept.

When Jim came to visit me, we put the modified painter's mask idea together and now needed to create something that produced airflow. I thought of a little device that could produce air pressure and could be designed in my workshop from the many pieces I had lying around. Jim and I managed to put an air valve together that allowed light air to flow into the mouth. We took it to the university to show Professor Sullivan to see if it would work.

It was a suitable method and he was able to experiment. I went to the university and tried out the system myself. There was a considerable amount of experimenting going on and in the meantime, I continued to modify the face mask to fit the air supply. Some time after, they were able to produce a suitable mask and it became the first in the world for sleep apnoea treatment. Jim helped Dr Sullivan to find a manufacturer and thousands have been made since. Professor Sullivan became quite famous for his discovery and came to dinner to thank me for my efforts.

## Tow Bar Ball Cover

A New Zealand company started up a branch division in NSW making tow bar balls and I was asked to make a plastic cover that could be removed easily, keeping the balls clean when not in use. I made a cup shape in polythene with a couple of ridges on the inside that gripped the ball, sitting perfectly overtop and preventing it from coming off, yet with a simple tug it would lift off.

So away we went and I made the mould for the cup and commenced to produce them in large numbers. Just when all was going well, the NSW branch division manager dropped dead and before you could say 'Jack Robinson' about half a dozen local manufacturers were producing the tow bar balls with a free cover, not made by me. The idea was not patented and so many other companies had copied my design. I still see them today, some fifty years later. However, it was not an entirely lost cause, as I did get some financial reward by making them to sell when the New Zealand manufacturing company went out of business. Unfortunately, I did become stuck with a large quantity of them. The free manufacturer offer didn't last too many years but by that time I had disposed of my stock and was in no mind to make any more. Finally, I came to my final innovative fiasco.

# Smiley - The Original Emoji (1971)

In 1971, an American magazine produced a cartoon cover image of US President Nixon and the leader of the Waterfront Union. On Nixon's lapel was a round badge with two eye dots and a broad grin to hide the union man's badge with a scour instead of a smile. I picked up my copy on a Friday afternoon and took it back to the factory to read the contents. That badge on Nixon's lapel got me thinking, *"Wow what a great idea,"* so I spent my time over the weekend planning how I was going to do this. First I would protect my idea and take out a patent on the smiley badge, copyrighting the name 'Smiley'. When Monday finally came around I was the first person outside the patent office in The Rocks, Sydney. I knew others would have the idea, so I assumed I'd be the first person at the counter that day. By the time the doors opened there were at least four other people behind me, one with the drawing of the smiley face. I beat him to it and I registered my design and copyrighted the name. By the weekend I was making the decos round, about 3mm thick with a small margin hole, and was negotiating with a printer to heat stamp the face. He had no idea how to make it so over the next weekend I made a printing plate and he was able to stamp the eyes and mouth with black plastic film. The original smiley was produced in the USA by Harvey Ball in 1963 however I produced the first smiley ever seen in Australia.

The Royal Easter Show in Sydney was about to get underway so I contacted many stallholders. They gave me a big order for Smiley badges with a length of cord and a jump ring for people to wear. We sold thousands upon thousands during the show.

Immediately after, I was contacted by a Sydney department store, Mark Foys. Before they got going, I had a visit from the Daily Mirror newspaper. They were keen to learn the story behind the 'smiley' and wanted to publish a full-page article with a model wearing lots of Smiley badges. Soon after Mark Foys contacted me, telling me they wanted to dress up all their vendor displays with huge Smiley badges, including setting up a special counter in the middle of the store, just to sell them. During this time I was approached by several companies compelling me to stop operating until I showed them my application. I informed them that if I stopped, they could pay my royalties.

I did lease out the original design to a candle manufacturer who produced a wide variety of candles using the face design. I then started to make other objects using the Smiley face. I was already making cuff links, coasters and t-shirts, so I produced a small number of button-sized Smiley faces as lapel badges. 2KO, a Newcastle radio station, ran a promotion called 'Stick on a Smiley'. I have a folder containing all these examples with a business registration certificate of the Smiley name dated 1971.

A man came to see me with a flock gun, which sprays suede-textiles flocking fibres onto an adhesive surface, so we coated the bottom of the reverse side of the Smiley badge with flock and sold these as coasters. Later I produced paper printed with adhesive on it so kids could decorate their school books. Mark Foys, after several months, said they had had enough. I was told the smiley was old hat. They took down the window display, closing the counter in the store. So by 1972, it was all over and no one wanted anything to do with it. I let my application lapse and got on to my next project.

It was still the early eighties when Smiley started to show its face again and by the nineties, they were everywhere. I googled the responsible company in England, to look up the

Smiley patent, only to learn that they were making millions in royalties.

## Plastic Clamp Toy Rings

Just before Christmas, around 1990, a customer called in to see me at work, trying to solve a problem he had after purchasing a shipment of toy sporting games from overseas ready to sell at Christmas time. He was told by customs they would not accept his shipment as the toy was considered dangerous to children. It consisted of a boxing glove tied to a rubber rope. Two combatants would try to box their opponents with the glove by pulling on a string. I suggested a safety stop on the cord. The customs officials accepted the idea and I told the customer I could make plastic clamp rings to suit. I started on a Friday and worked throughout the entire weekend, with little sleep, and set them up in a moulding machine. First thing on Monday he had a supply for his staff to assemble the rings.

### *All in All...*

I had fun making and designing things in my life but never really capitalised on them with patents and receiving royalties.

I was great at manufacturing new ideas but didn't have the foresight to see the potential of some of the things I made. There were many other ideas but most stayed as drawings in the bottom drawer of my desk.

# CHAPTER 13

# Bruny Island (1996-2020)

In 1996 I was going through some family records, uncovering my connections with the pioneers who went to Tasmania in the early 1800s. This gave me an idea, and I suggested to Marian that it could be a good idea if we moved down to Tasmania to get away from this dreadful urban life in Redfern.

Marian and I took a trip down to Tasmania and drove all over the island, including Smithton, where her family lived. They were timber contractors. We decided, yes, this was where we wanted to live in our retirement years. We both loved the idea. We planned to sell our Redfern home, moving to Tasmania sometime during 1997 in my 66th year.

## My Vision Loss

On my first property search trip down to Tasmania, I was looking at various places around the Hobart region searching for an old house with plenty of grounds. However, the top of my 'to do list' was to visit Port Arthur on the Tasman Peninsular. I wanted to stay in an old convict hotel in a nearby town, Nubeena, on the way and booked for one night.

There I woke up to a nice sunny day but noticed there was a black line across the sight in my right eye. I recognised this as it had happened before. Back in 1980, I walked into a log, banged my head and a similar black line came up my vision. It was diagnosed as retina loss. Back then I went to hospital for the retina to be repositioned through eye surgery.

Hoping that it might just go away, I continued with my plan, visiting Port Arthur. I looked at the ruins and spent quite some time in the museum, taking a great interest in the padlocks and locks. On my way back to Hobart it started to rain. Now I couldn't see out of my eye and all I could remember was worrying about the nuisance of pouring rain and my lack of sight.

I drove straight to the airport, thinking I must get back to Sydney to get my eye fixed, but the man behind the counter informed me there were no planes until tomorrow, 8am to be exact. He inferred that the flight was possibly booked out so I had to wait until the morning to book a seat. I was losing my patience, telling him about my eye and that I couldn't drive any further, nor did I have a hotel for the night. I was considering sleeping in my car but I didn't want to go outside as it was bucketing down with rain.

I made myself as comfortable as possible in the airport lounge when I was approached by the security guard around 9pm. He wanted me to leave because the airport was closed. I said, *"Sorry, I'm not going to move, I'm going blind in one eye and will not go out in that rain."* He advised me that he didn't want to find me upon his return after securing the building. I was determined to lie down and stay put. He came back and after more discussion, he compassionately presented me with a blanket.

The next morning, I was given a seat on a flight that left within one hour. I informed the car rental company that they now had to pick up my car at the airport. Having already spoken to Marian on the phone, she had organised an ambulance to meet me at the airport in Mascot, Sydney, upon my arrival. I was whisked away by the ambulance, with my eye completely covered, to the Royal Prince Alfred Hospital. The eye doctor informed me that it just so happened they had a visiting Japanese eye specialist that day. The first doctor gave my prognosis a low chance of success with surgery but the Japanese doctor gave me a 50/50 chance to get my eyesight back. He operated, and then I had bandages covering my eye from Monday morning until Friday, before the reveal.

I managed to get through the week, in nervous anticipation. I had resigned myself to the fact that I could probably get by with only one eye. The doctor said it was touch and go, describing to me that it was like marmalade in there, with bits and pieces floating around trying to manoeuvre the retina back into position. He used a laser gun, and this was a first for Australia, of such an operation. He could see that my prior operation had been supported with rivets but they had collapsed. The bandages finally came off, and I could see a bit of light. He put his hand in front of my face, asking, *"How many fingers?"* I counted one, two, three. The surgery was a success and he said, *"I don't believe you know how lucky you are."*

Later I was having my eyes checked by another doctor in Sydney for glasses, and as it turned out, he was a student of the above Sydney specialist. He had gone on to open an eye hospital in Tasmania, becoming quite successful. I remember he had the most beautiful deep voice that I will never forget. One day, as I was living on Bruny Island, I listened to the ABC radio and heard his incredible voice again. The specialist was being interviewed by Margaret Throsby on the radio. I immediately rang the optometrist, to tell him the eye specialist was on the radio right now. He was on the golf course when I rang and he already knew because he also had the radio with him in his golf buggy.

To this day, I have had no problems with my retina. I do use glasses for driving but can see perfectly.

### *Back to Relocating*

Instead of buying an old house around Hobart with land, we decided to look for an empty block to build a new house. Finally, I was in a real estate agent's office in Hobart, having a few recommendations, and just about to walk out the door, when a voice from the end office called out, *"I have an interesting block of land that has just come onto the market, on Bruny Island down at Cloudy Bay that the customer might be interested in. Here's a brochure with a picture of it."* I did see other blocks on the mainland but then took a trip across to Bruny Island by ferry to have a look. I was thinking, it was nice of this lady to recommend the land but I did feel it was a bit too far, being well out of the way. But at least I would be polite and have a look for her.

Arriving on the island I drove for about 45 minutes to pick up an agent from Lunawanna. We travelled through some very open countryside, and climbed up a hill, going along a very gruesome looking bush track. By then I had already decided this was not it. But at the top of the hill I parked and got out of the car, and my instant reaction was, *"I will take this one."* It was a very big heartfelt YES.

The land had a broad beach in front of where we would situate the house, high above the beautiful sands of Cloudy Bay with outstanding ocean views. It overlooked the Tasman Ocean, being a large block of 50 acres. The neighbours on either side were nearly a kilometre away from where we would build, so you couldn't see them. It was a perfect spot. It didn't take me very long to get back into Hobart with the feeling this block wouldn't last too long. I paid a deposit immediately and thankfully I did because many people were chasing it.

Now back in Sydney, I had to find an architect to design the perfect house for this very special block. I had a magazine on architecture and noticed a particular building firm featured in it, whose work I admired. I phoned that company, asking if I could speak to their architect. As I was building in Tasmania the building company excluded me from speaking to the architect. However, shortly after hanging up the phone, they rang back, apologising, saying they couldn't stop me from talking to their architect to design a house. I was most pleased.

First of all, we agreed to his commission of 3% of the total cost of the house. I thought this was a fair consideration. He made us beautifully illustrated drawings and a set of plans which we approved, disapproved and analysed carefully until we finally came to a firm decision on the design. The architect helped us find a Tasmanian builder, and said he could build for us on Bruny Island but had to wait until we got down there, which was soon after.

## For Sale: 56 Pitt Street, Redfern

*"Circa 1896 Nizam" is a fine example of the Victorian era. Unique in its size and ballroom facility this lovely old terrace exudes a charm and warmth that is rarely found so close to the centre of Sydney. From the ornately moulded cornice in the reception rooms to the marble fireplaces and marble-panelled bathroom this wonderful six bedroom residence will transport you to a time and lifestyle that is rapidly disappearing. On 569.2sqm of land with a Residential 2b (2) zoning this fabulous property would suit a large family, home and business or subdivision."*

By now we were excited to sell our house in Redfern. The timing was good and it sold for a very good sum of money. This helped us fund our new home project and put a few dollars in the bank. The cost of moving was quite monumental as we had accumulated some very fine antique furniture thanks to my business of restorations, not to mention my collections and the items in my workshop. Marian also had her own fine collection of beautiful linen, sheets and tablecloths. The carriers managed to fill nine containers for us to take down to Tasmania.

We also had four dogs; a German Shepherd 'King', Saint

Bernard 'Alice', Belgium Shepherd 'Kaiser', Chow Chow 'Earnest' and three cats to take with us. While we were planning our retirement relocation there was a company advertising the sale of miniature pigs at the Sydney Showground. We thought, wouldn't it be nice to have a little pet pig to take with us, joining our large family of pets? We bought her for about $100 and named her Hermione after the English actress, Hermione Gingold (1897-1987), who I had admired throughout the years. Our beloved pig grew to an enormous size. Before she died, she weighed over 300 kilos – definitely not a miniature. However, upon losing two of our dogs, King and Alice, after a long full life, Hermione died soon after in 2017, as they were her buddies.

Our previous neighbour Aminta had a flying school and a tugging service to pull gliders out of the Bankstown airport. Travelling to Tasmania, from Bankstown airport, we had all the pets and myself, plus my wife, her sister and my brother-in-law, who also wanted to have a look at our new land. I chartered an aircraft for us all to travel down together. Upon meeting the pilot at the airport, he apologised that the plane requested was unavailable. So he had to get a smaller plane for my wife, her sister and husband plus the cats, leaving me at Bankstown with the four dogs and our pig, to wait until he came back.

Unfortunately, the plane was not in good condition when it arrived back to Bankstown so we would have to get another one, two days later.

I was kindly offered to stay at the pilot's house. The pilot said, *"Don't worry about the fence, the dogs can't get out."* But I went to feed them later and they were gone. There was a gap in the fence. Thankfully I found the dogs; they had only got down to the river and were happily chasing birds. I managed to get them back to the house, pulling them back through the hole, then sought to repairing the fence. I had kept the pig in her cage the entire time and she didn't like it. When the pilot finally arrived back and we could board our flight, he said, *"I have a feeling there is something little amiss here about moving animals from one state to another. I tell you what, I will engage with the man at the gate, offering to have a beer with him, while you drive the car to the hanger to load the animals onto the plane."* It took three people to lift the pig onto the plane in her cage, then the four dogs in their cages. A second pilot was sitting up front so I had to sit cramped up in the corner, down the back in-between the dogs. It was a very slow flight down. We arrived at 6am.

The second pilot disembarked and my pilot went about filing his flight plan. He said, *"The fellows in the shed will help you unload your animals."* My brother-in-law had arranged a truck to be there from a hire company. I went over to the shed and although the truck was there, no one was there to help me. I drove the truck over to the plane and managed somehow to get those animals off and into the truck all by myself. I still had another two dogs to move off the plane when this nosey fellow with an American accent said, *"If the dawg gets out, I will shoot him."* I said, *"They are alright,"* even as I noticed one of the dogs had chewed a hole in the plywood cage and was just about to get out. I shoved the last cage into the truck. Then I manoeuvred the pig out. It was a hell of an effort, lifting the cage then pushing it onto the truck, but I was determined to get it done. The airport office didn't open until 9am but I noticed another truck coming out of the warehouse, so I immediately jumped into the van, sneaking out behind this truck before the security guard could stop me. Not looking back, I belted down to Bruny Island.

I arrived three days later than the others. We didn't have mobile phones or landlines to notify of any changes to my travelling ordeal. Everyone had been terribly worried but it turned out well in the end and they were thrilled to see me. We had rented a house to live in while we built. I had arranged for the house to be fully fenced so we could let the animals out of their cages.

Unfortunately, there was bracken (large coarse fern plants), and the pig did a wonderful job cleaning it up however she passed out from eating it. I thought, *"Oh my goodness, there are no vets on the island."* But I managed to get one to travel out. She was very understanding as to why we had the pic. She inoculated Hermione, leaving me with a syringe, and asked for me to dose her up in four hours. She was out to it, lying on the ground, and as instructed I snuck up and pushed the needle into her bottom. She jumped up, running around the yard, with the syringe sticking out of her backside. I did finally manage to squeeze the serum into her. She was fine for the rest of her healthy happy days of at least eighteen years.

My next job was to get the builders over to prepare the ground with their bulldozers and set the foundations. The timbers arrived and the carpenters put up the frames. We had the house clad in corrugated iron, which I thought would be ideal for protection from the strong winds and rain of the area. After about 8 months, we had finally finished the house and were able to organise the rest of our containers to be delivered. They were stored alongside the house, down on the lower side, at the back where I had a very big shed built. We were able to load all the containers into this shed.

We had rented for almost a year as it took that long to build our new house. In negotiation with the owners, who were anxious to sell, we agreed to buy the house. I knew when I went on to sell it, I would get my money back.

Finally, we filled the new house with our antique furniture and numerous collections, and it became home.

# First Night in Redfern

It was late 1997, our first night in 'Redfern' – which we had called the property – and we found ourselves surrounded by all our treasures. Enjoying the company of our dogs, cats and pig, we felt absolutely delighted to be in our new home at 884 Cloudy Bay. The 500-metre road that led up towards our property didn't have a name, so we took the liberty of naming it 'Terry Street', and placing a sign on the power pole.

## *The Guided Tour*

As you walked along the concealed driveway, up to the portico, you would find the front door had 'Redfern' engraved into the glass. The portico had a pedestal of birch with four timber columns that held up the roof. These columns came from Redfern's oldest house in Pitt Street, 'Dragonfly', opposite our old house. Dragonfly was demolished in 1996 by a developer who built three terrace houses and we saved the columns.

The grand front door, made from cedar, vintage 1890, opened from the eastern side of the home, which faced the stunning vista of Cloudy Bay. This beautifully hand-crafted door featured a brass knob in the middle, with an elaborate lock.

The iron hinges added an impressive entrance to the castle. Surrounding the door were glass panels on either side and above, to admire the beach below. At the base were two dolphin figurine doorstops.

Upon entering the hallway you walked into a three-metre wide entrance meeting double doors to guide you into the conservatory. We built the conservatory in 2005, some eight years after moving in. As you stepped into the conservatory your eyes would fall upon the hand-crafted table that I designed. It had a circular marble tabletop within a brass frame sitting on brass legs. I placed a chandelier over the central position of the table. This came from our house in Sydney and featured five brass balls.

As one might expect there were a lot of plants in the conservatory and we utilised a lovely display cabinet from Indonesia for the plants. I bought the damaged piece at a very good price from the Balinese House in Hobart because I was able to repair it myself. It was made of wrought iron, having doors and open shelves. Upon every joint, there was a little cast iron symbol, for example, a dog or a monkey. It was an ideal piece for displaying our pot plants. Also, in the conservatory, we had many large decorative ceramic elephants. It was a wonderful place to enjoy a glass of champagne at the end of the day, watching a glorious sunset shift into the night sky.

Just outside of the French doors, in the hallway, in line with the doors onto the libraries was a wonderful seven-day large grandfather chiming clock. It was crafted in England by a clockmaker in the midlands, circa 1780. I enjoyed the weekly ritual of winding up the clock every Friday night, right up to the day I left this house, hoping the new owner would enjoy it too.

### *35,000 books filled our two libraries*

Half way down the hallway a door opened into my library. The door was made from old red gum, being an old standard government issue office door. It had a lock escutcheon, brass handle, brass door knob and a piece of brass push plate that was fitted with an iron lock, which fitted on the edge of the door. As you can probably tell by now, I wanted old world doors of elaborately fitted width. Past this door, my library was fully lined with bookcases from floor to ceiling and had two large, long windows that overlooked the bay. A large Persian rug covered the floor, with my billiard table sitting central upon it. The blue and red rug was handwoven in Pakistan, being a Princess Bokhara original. A marble fireplace was at the other end of the room. Throughout our house we had beautiful rugs. In our bedroom we had small rugs made from silk and lambswool, known as QUM fine Persian rugs. The silk rugs were made during the 1920s by Armenian weavers.

## *Records*

In my library I also had three rows of the 78 RPM record collection. This included some original Bakelite records made out of hard synthetic plastic. Leo Baekeland (1863-1944), a Belgian chemist and entrepreneur, discovered how to make the first synthetic plastic in 1907. Thomas Edison (1847-1931) along with Jonas Aylsworth (1868-1916) combined Baekeland's plastic with a newly discovered technology of their own, using various compounds, ultimately producing the hard plastic records.

We enjoyed playing recorded music on our big hi-fi system which sat in one corner of the Breakfast Room. It consisted of two large cased speakers and a very handsome multi-player with three speeds, 78-45-33 RPM, along with a good quality amplifier.

Yes, Marian had her library too. She had a comfortable chair dating back to the 16th or 17th century. It had carved eagle arm-rests, with carving inlays on both the seat and the back. I believe it was worth a small fortune, however, I never got around to fully restoring it. I bought the chair from a dealer who thought it was much too big a job to restore, so I paid only a small amount for it. Someday someone will find a good use for this incredible throne chair. Marian also had an 18th century sofa covered with a beautiful tapestry design.

Marian's library had a marble fireplace too. I fitted both the marble fireplace and marble flooring surrounding it shortly after the house was built. A Georgian desk was beside the fireplace with a French marble mantle clock on top. The clock sat inside a large lead-light glass case made by an old friend on the island. But it was a complete disaster when I wanted to sell it. It was a very rare piece and I was planning to make a lot of money. I took the clock apart, taking the pendulum and key out, and placing them carefully onto the pedestal. The movers took the clock to the auction room, however, they didn't take the pendulum out of the box. Intact the clock would have reached $15,000 however, without the pendulum it was sold for only $1,000.

The entire house had high ceilings with exposed support beams of 120mm x 600mm. Over my doorways we installed plinths, to house the beautiful ornaments above the doorway entrances. The library entrance plinth displayed an apprentice piece, a miniature wardrobe that was beautifully built with a miniature coat hanger, doors and draws. The craftsmanship was magnificent. Such apprentice pieces were originally built for salesmen to take around to show prospective buyers the company's furniture designs.

On another plinth, we displayed a large figurine, being a set of two little girls. We acquired them because they reminded us of my two nieces, Erica and Jane. The figurines were left in the house upon the sale of the property.

In the hallway we placed a 100 year old pianola and a piano from Stewart Symonds. The first piano made in Australia was a copy of this one. Stewart Symonds had a collection of about 180 pianos and he bequeathed his entire collection to the Perth University before he died. However, I had bought mine from him many years prior. The pianola had a pedal foot pump and was designed to play piano rolls but required a separate piano for the entire experience. To operate it we wheeled up the piano to the pianola, so the little fingers from the pianola could play the piano.

In the hallway, opposite one of the pianos, we had our elephant collection displayed in a glass-fronted sideboard. I exhibited all my best ebony and ivory elephants. This collection was given to my daughter, Belinda.

Directly ahead was a pair of French doors I acquired from an old church in Glebe that was being demolished. The glass panels were decorated with religious symbols however when I fitted these into the Pitt Street House, in Redfern, I broke one of the panels, so I replaced them with etched panels of a vase of flowers.

Throughout our home, we laid Tasmanian oak flooring. This can be a little misleading because Tasmanian oak is a hardwood produced from three species of eucalyptus trees. The only exception to the Tasmania oak flooring was in the wine cellar that was just off the dining room. It had a wonderful jarrah parquetry floor that we had kept in storage for quite some time. The original parquetry came from a men's clothing store that was closing down in George Street, Sydney.

As I knew the owner, we were invited to help ourselves before the entire building was demolished in 1980. After I laid the jarrah parquetry and then sanded it, we oiled the flooring with linseed oil and oiled it again six months later. The flooring only needed an easy wash to bring up the lovely grains of the wood. I did enjoy this project.

Walking through the library, another hallway went off in both directions, left and right, with the dining room directly in front. To the right was the kitchen and breakfast room. To the left was the toilet, including a urinal. The plumber was a little taken aback when he discovered I wanted to install one, as he believed it was not permitted. I sent an application off to the council for a by-law to accept it, but I never knew if it was accepted. I could only assume it was, as it was installed effectively by the plumber who fitted extra pipes to accommodate it. I must say, when visiting gentlemen went to use the toilet, they were most impressed, enjoying the ease of such a device. In the toilet room, there was a lovely old hand basin and an old Victorian toilet with a wooden seat.

## *27 doors*

Every door in the house was salvaged from old buildings, demolished houses or from the local tip. The doors had been packed neatly into a container after I had bought and restored them. In this container was a total of 27 doors. For example, the French doors to the breakfast room were originally in the old Post Office in Warwick, Queensland.

When we arrived in Tasmania, I was offered a large quantity of locally cut Blackwood, which had been stored in a shed for about a hundred years but never used. I bought the lot and used some to make a table and other projects as they came to hand.

We spent much time, talking, reading and entertaining friends sitting around the breakfast room fireplace. It was a focal point during winter days. The door into the breakfast room was made from a combination of sequoia wood and American redwood, the world's tallest trees. Also in the breakfast room as a central feature was a five-light chandelier I had made in Redfern out of brass fittings. I hung wall-to-wall pictures of the Sydney Harbour Bridge. Marian's sentimental sofa wasn't the prettiest piece of furniture but was extremely comfortable to sit on and double as a daybed, to enjoy a granny nap in the afternoon.

Upon entering the dining room, through another set of French doors, you would find an Andre Charles Boulle Commode, a prestigious piece of furniture that was designed in the late 17th century for the King of France Louis XIV, also known as the Sun King. It was an original piece of furniture bought out to Australia by Sydney Snow (1887-1958) for his department store, Snow's Emporium, in Pitt Street, Sydney. In 1930 he donated the commode to the Museum of Applied Arts where it was displayed on the staircase between the ground and first floor. When the museum closed down, the Power House opened and this is where the French Boulle Commode was sold to us at auction. The commode was fitted with goat's feet and brass castings for the legs. There were black cast iron blacksmith locks with five drawers. I did not repair or alter the commode and sold it at auction in 2020 for $40,000. I shared this money with my family.

Next to the commode was an early cedar sideboard, carrying our 18th century sterling silver tableware: silver serviette rings, silver trays, silver breadcrumb trays and other dining tableware along with linen napkins.

There were two corner marble tables in the dining room with 18th century lamps. These lamps were not working but they were in their original condition. The authentic white Carrara marble fireplace was complete with fancy columns and a mantle. The fireplace grate was supported by two large framed brown dolphin figurines. I laid a marble surround to this fireplace. It was a replica of one in a hotel in Hobart, near Salamanca Place, and we used it quite often.

We placed a lovely 18th century mirror above the marble mantle, and a huge candle chandelier hung from the ceiling. This piece had been acquired some time ago. I couldn't assemble it before the auction, so the owners gave it to me for free, however, it was left behind with the house when I sold it in 2019. This piece was gold plated; the rings supported 24 candles and it was suspended from the ceiling with eight sets of chains.

Many years ago, I was given an 18th century piece and asked to restore its pair of ornate cast iron legs. I removed the rust to expose the bare metal, which I oiled. Much to the client's surprise, I asked if would he mind if I had them copied, and he was willing to let me do so. At the foundry, I was convinced to have them copied in bronze, making them unique. Upon completion, I kept them for many years waiting for the chance to use them to finally build a large table in the breakfast room.

Further down the hallway was another big room that Marian used as her fabric room. She loved sewing and had two sewing machines, one being a Singer and a portable one from Switzerland, an Elsa Vintage. In her fabric room, she had six wardrobes and two cedar chests of drawers, filled with her extensive collections of linen. Some of the beautiful lace and linen dates back to the 16$^{th}$ and 17$^{th}$ centuries. I have no idea about the value of her collection but I did give some of Marian's linen to her sister when she passed away, leaving most of the linen in the house for the new buyer.

She had every inch of wall space displaying a picture. This set the stage for the rest of the house. We displayed over 800 original oils, watercolours and a few prints, some photos and some family pictures. I left about 700 behind but brought about 100 pieces with me to Queensland.

Down the hallway towards Marian's Fabric Room was a powder room. In here the toilet was made by Thomas Crapper (1836-1910), the engineer who refined the flushing toilet model during the 1860s. In this room, we had wall-to-wall pictures of Tasmania, including a panoramic view of Hobart. Another door from the kitchen had coloured glass that led onto the timbered walkway, which I built around the entire corner of the building, from the laundry to the breakfast room.

The wine cellar had wine racks from floor to waist height, with a toasting table above and a bench-mounted corkscrew. On the right side of the cellar, I displayed my collection of antique corkscrews. Above the wine racks, up to the ceiling, we displayed our glass collections. On the left side were six shelves displaying a myriad of old and mixed glasses.

We had a rare sherry glass with a fluted spiral base maybe worth over $1,000. There were 19$^{th}$ century champagne glasses and 17$^{th}$ century wine glasses. We had a Prince of Edinburgh engraved glass, as a thank you from the British Trade Export company for boosting exports by British Oxygen Company (BOC) who had introduced fluoride into aerosol cans, for hairspray, in Australia.

Helena Rubinstein's company was the first company to endorse fluoride in their hairspray products. A cosmetic side story, in 1978, I made lipstick cases for Elizabeth Arden.

## *$5 Penfolds Grange Hermitage*

The low shelves held a large variety of vintage wine from Marian's collection and a few of mine including several Penfolds Grange Hermitage, dated with typed and written labels, well before the time they were printed. I sold them at auction, expecting $1000/per bottle, but sold them for only $100/per bottle. Mind you I had bought them for nothing, at a hotel in Gosford, many years before. The owner was emptying his cellar and found these Penfolds, advising me, "*You can have them for $5 each,*" so I bought half a dozen, we drank a couple and took the rest to Tasmania.

Into the kitchen, we had an AGA four cooker oven, made in England, running 24 hours every day on heating oil. Initially, it was half the price of diesel and very cheap to run. As time went by the cooking oil became a highly taxable item, the same as diesel. However, we continued and just had to pay the higher costs of running our stove. The kitchen had a lovely walk-in butler's pantry next to the stove. My first job was to build all the shelves from floor to ceiling to hold everything. We stored food but also had a large collection of old kitchen memorabilia in the pantry. There was an old apple peeler, a clockwork rotisserie and a sausage grinder, an old demijohn ceramic container, old mixing bowls and rolling pins.

In the centre of the kitchen we installed a government-built school physics bench from the 1920s with a porcelain sink and a spartan water system. There were holes in the bench for the classroom Bunsen burners, which I filled in. There were four large drawers and below these were cupboards, so we had shelves for many things. There was also a large glass door dresser for glassware. Two big drawers held our 18$^{th}$ century silver tableware. Shelves below held pots and pans with a door to protect them. On the next wall was another dresser, with open shelves. On the very top shelf of both of these dressers, we showed off a lot of old copper water heaters and a tea urn.

Opposite these dressers, was a marble-topped table. Both this table and the physics bench were used to prepare food. Marian and I loved to cook for our friends. Off the kitchen and doorway to the dining room was a scullery. Here I would fill up the large stainless steel laundry tub with a drying shelf, to handle the mountain of washing up that was required after entertaining guests.

There was a door behind the kitchen sink leading onto the

deck that ran along both sides of the dining room, kitchen and breakfast room. On the outside was one of a pair of underground water tanks holding approximately 100,000 litres. The reason for so much water in reserve was in case of bush fires. I had in the shed a petrol pump suitable for the job. When we did have a massive fire, the local fire brigade had sufficient water to handle the situation. In my twenty-five-odd years living there, the water never went below about 100cm deep in the tanks. However, it took two years of rain before they filled up completely. We would wash our cars weekly due to the dirt roads we lived on, plus with watering our gardens, we always needed plenty of water. In contrast, since living in Queensland for four years now, I have never washed my car, because of the rain.

### *Fishing at Cloudy Bay*

My land at Cloudy Bay adjoined the Cloudy Bay Lagoon. The lagoon only had a small entrance, at the end of our peninsula, allowing free-flowing water in and out. In my early days living there I had a friend and we would go fishing at this outlet but from the other side. Our idea was to fish on the outgoing tide. We would stand on a rock simply reeling in the fish: bait the hook, throw in the line, pull out, remove the fish, bait the hook and repeat. One day we managed to catch mostly flatheads and leather jackets. We moved back from the water's edge and I got a nice fire going, grilling the fish. We ate the freshly caught fish and drank a few cold beers. I was in my bliss, a most joyful time.

On my side of the lagoon, there was an expansive long beach with white-shaded seashells. At the end of the beach the community had a large water tank to be used as a washing place for the shells. People from the island, including the father of Malcolm 'Butch' Barnett (who ran the oldest timber mill on the island) would drive past my place, down to the beach. We would engage in washing the shells and bag them up for loading onto the back of a truck. Following this exercise they would drive into Hobart with their shells to sell to the only Tasmanian glass works, now long gone, but I did provide labour and glass products to the wealth of Bruny Island.

On the other end of Bruny, near Robert's Point, there is a beach home to fields of saltbush growing wild. Evidence on the beach is still there after the saltbush harvesting in the 1880s. The saltbush was boiled, and then the liquid dried out, the residue being salt crystals. This salt was then sold to the soap manufacturers of Hobart. The salt business's early papers were given to the national trust and can be perused at a nearby market, at the Soap Factory in Hobart. It is no longer operating but Bruny did play its part in the colony during the 1880s. Making soap was overshadowed by the vast timber industry that thrived on Bruny Island. It had over twenty mills at one time exporting timbers to the rest of the world.

### *The White Wallaby*

Bruny Island is home to a sizeable population of white wallabies accordingly, they are not considered an endangered species and I would often have them visiting my property. One day I managed to get a photograph of one to prove they existed. I enjoyed seeing them on my land and they soon became quite tame, not feeling threatened with my presence.

Bruny Island white wallaby on my land at Redfern (2010).

## The Bruny Island Men's Shed

My so-called leader's life came to light again 10 years after our move to Bruny Island, when I formed the Public Library and stayed on as the secretary, though only for a short while. I then founded the Bruny Island Men's Shed. I took on the role of president, which I held for nearly nine years. I must say it was a very pleasing activity. I enjoyed gathering more and more men to join us. We undertook many jobs in the community. The Men's Shed grew from our first meeting in an old health centre's morgue to having our own big shed on land that was donated to us by the Kingborough Council.

Me and Will Hodgman, the Tasmanian Premier, on the opening day of the new Men's Shed (2016).

BRUNY NEWS – BRUNY ISLAND MEN'S SHED NEWS – PUBLICATION BRUNY NEWS, JUNE 2016, PAGE 9, MAD HATTERS

"The ceremony of turning the first sod for the new Bruny Island Men's Shed took place between showers at the chosen site on council land at Alonnah on Monday 9th May at 11am. The two venerable gentlemen dressed as appropriately as one does on these occasions are (as shown in photo): the grey topper our current vice president, John Davis and in the other corner in daggy trousers was yours truly, performing the act of turning a sod of earth. As befitting the *duties of president of the Bruny Island Men's Shed, and witnessed by a number of our members while photographed by John Pforr. We did not advertise the function because we were advised that the council were about to level the site and if we wanted to signify the occasion, do it now, so we did. Stand by for a completion day and this will be advertised. A suitable dignitary will be chosen to cut the ribbon which will be followed by much frivolity, wine and good food. Following the levelling of the site, with thanks to Kingborough Council and their merry men, we will attempt to get a concrete slab laid and this will be followed by the shed contractor to assemble the bits and pieces. When? Before Christmas I hope, well you know how these things take their time. More information next month, Terry Butcher, president BIMS."*

## An Original Chainsaw

I acquired the first chainsaw that was used on Bruny Island. It was made for the timber industry, having a very heavy, world-renowned, Villiers' engine. It sold for only $10; it seemed the historical significance was not appreciated.

*THE LOST ART OF PIT SAWING - PUBLICATION*
*KINGBOROUGH CHRONICLE - 6 MARCH 2012*

Terry Butcher of Bruny Island shares with readers his fascination with pit saws and other long gone wood working tools. *"There are thousands of collectors, collecting thousands of items for a thousand different reasons. I am one of those collectors, and my reason for collecting is to identify the item in its place in society, its purpose, age, and at what period of time it was useful."*

I'm interested in collecting pit saws. They were used in a saw pit, where the top sawyer would grasp the handle or tiller and, following an inscribed line on top of the three log guides, he would saw towards himself as he walked backwards along the log.

The bottom sawyer would control the saw on the bottom and with a handle that is easily detached, it would be necessary to move past the small supporting logs, and in order to not cut through them, the log was moved back a little and the saw remarked to continue the cut. All logs were cut lengthways in this manner before the invention of the circular and band saws, about the middle of the 19th century. Pit saws have been around for a long time but are rarely seen these days, apart from in the forests of Brazil and some of the Pacific Islands where they are still being used. Many here were converted to cross cut saws with the tooth front being changed and a different style of handle fitted.

Now we come to the question as to why I wanted a pit saw in my collection. I have been collecting for many years wood working tools for all phases of wood usage but when I came to live on Bruny Island, I became fascinated at the long history of timber getting throughout Tasmania. I soon delved into some study and discovered Captain Bligh (former Governor of NSW and infamous for the mutiny on his ship the *Bounty*) had beached his vessel in a creek at Adventure Bay, and sent off the ship's carpenter to locate a tree nearby that they could convert into planking to make repairs for his vessel, before they left on Captain Bligh's last trip.

The trees in this area were large, straight and very tall. I imagine after felling the tree he chose a typical felling axe and cross cut a log into planks. And so in 1792, here for the first time an industrial event taking place, possibly for the whole of Tasmania.

I have a selection of felling axes, similar to what would have been used, as well as several clamping dogs that hold the log steady, and cross cut saws. Mine are of the mid and late 19th century and also the 20th century. Now I have finally acquired the pit saw to complete the set that typifies this first and momentous event. Other tools used in the preparation of timber used for a multiple of uses – railway sleepers, mine pros, housing and bridges – were side axes, in cases of large converted logs to a square or rectangular shape that were not cut by pit saws but normally shaped using side axes.

The timber we use today comes through to us from mills at supersonic speeds and ready to use, while back in those days, all that work was done by hand; skills now no longer needed and forgotten, but at least we have the old tools they used and the knowledge of how they were used.

# The Bruny Lifestyle

## *Radio Waves Research for NASA*

We had a regular visitor, William Erickson (1930-2015) who was an astronomer. He lived not far from our place and he had a small radio telescope. When he and his wife, Hilary, studied radio waves from the sun they reported their findings regularly to NASA, who employed them for their research. Bernard Mills (1920-2011), whom I knew from the Mill Cross Radio Telescope days, was the Professor of the Astrophysics Department at the University and often met up with William on Bruny Island. We were all good friends and enjoyed socialising together.

Second in charge of that department was Dr Charles Munro, alongside myself, being third in charge. He was pioneering radio astronomy but finished up living in Tasmania because of the hole in the atmosphere where he could get a better signal. Charles came every year to visit William and his wife. He had died shortly before we left Sydney for Tasmania. Oh, what a loss, it would have been wonderful to have him visit us at Cloudy Bay. William was the first person to be awarded the prestigious Grote Reber Medal, for his lifetime of contributions to radio astronomy. I was present at the University of Tasmania when he was awarded this medal and have a photograph of that day. Grote Reber (1911-2002) built the first big dish in Tasmania to map the skies and measure frequencies, building on the work of Karl Jansky (1905-1950) who discovered radio waves in 1933.

### *The Chinese Veneer Mill*

Dick (Richard Geeves), my very good friend, rang me one day and said he had been asked to check out a new Chinese veneer mill in central Tasmania. He asked if I would like to join him. Marian agreed for me to go as it was something I knew very little about the process although I had studied peeling machines. In central Tasmania, dominated by both new and old forests, was a huge factory devoted to cutting logs into veneer to be exported to China, then to be converted to plywood sheets and sold back to Australia to be used in our building trade.

Visitors were not welcome but an exception was made for Dick representing the Tasmanian Timber Industry. A condition made by the Australian and Tasmanian Governments was that all the veneer mill employees had to be Australian citizens, and only management could be Chinese. On arrival, one came across a 3-metre-high white fence, and an employee moved the barrier gate. Dick showed his invitation and we were let in. Inside we were met by a young man to show us around and answer any questions we may have. The first thing we noticed was a mountain of logs of dressed timber cut to a fixed length of exactly 4 metres long and approximately 3 to 400cm in diameter. All the logs were fully stripped of bark. We then found ourselves entering a huge drying room. The veneer, as we were to see later being cut, was first steamed and piled into stacks, strapped together and ready for export.

Dick's parents were honoured to have a town named after them, Geeveston, which is located 62km southwest of Hobart. The area is well-known for its forestry heritage and apple tree farming.

Exert from Richard Geeves' life story, published by his family

"In the early 90s, the Forest and Heritage Centre was contemplating the purchase of Vince Smith collections of antique manually operated wood working tools. I was on the Board then and offered to get an expert appraisal. Henry Black in Sydney said, he would come down but told me of a Tasmanian resident who would be just as good. He lived at some unknown place called Cloudy Bay. That is how I came to meet a most delightful and unusual couple as Terry *and Marian Butcher. We have become very good friends and have done some unusual and offbeat things together. I could write a book about them alone, but will only touch on a few things. Amongst other things, he is an avid collector of anything: Sydney Harbour Bridge paraphernalia, Cork Screws, Wood Working Tools and Books. One day I wanted to give him a prescription for Colchicine but could not remember the size of the tablet. Not a problem. "The right-hand end of the third shelf on the left side of the doorway going towards the front door in the drawing room". I had been directed to not one but two full copies of the British Pharmacopeia.*

*For the last three years Marian's sister, Grete, has come over and stayed with Marian and Terry and I spend three days visiting various remote areas of Tassie. He is very good company and full of new revelations. They both have a high sense of community responsibility and are always doing little things for others. We were lucky to have met them."*

Me and Richard Geeves

*Pennicott Wilderness Tours*

It was Bob Pennicott's idea to take people on tours around Bruny Island collecting lobsters for the seafood restaurant. He started by buying a converted rubber ducky taking passengers for tours around Bruny Island, which has been highly popular. Today his tours are on a much grander scale. I took a lot of trips with Bob on the boat, taking us around the Friars and three or four islands at the bottom of Bruny Island. The abundant ocean wildlife includes colonies of seals, albatross and other sea birds. The natural beauty of Bruny Island captivates anyone who visits.

### MEET THE NEIGHBOURS - AUSTRALIAN GEOGRAPHIC - PUBLICATION
### AUSTRALIAN GEOGRAPHIC, JUL/SEPT 2005, ISSUE 79, PAGE 98.

"Cruise operator, Bob Pennicott and his artist wife, Michaye, live at the entrance to a placid lagoon separated by a sandbar from Cloudy Bay, on Bruny's southern coastline. Their house looks east through a screen of gums across Cloudy Beaches, Bruny's most popular surf beach and the Southern Ocean lies beyond...The Pennicotts have neighbours but they aren't visible from their house. Terry and Marian Butcher moved from Sydney to live in a large, prominent house set in a commanding position just back from Whalebone Point - a small headland in Cloudy Bay. Nobody seems to watch TV, preferring the sound of the ocean intermingled occasionally with classical music. They have time for some serious hobbies. Terry's antique tool collection, for example, is one of Australia's largest and includes an amazing 1500 corkscrews, the oldest dating back to 1750... It seems that to blend into the Cloudy Bay community you should be a remarkable individual who prefers personal reality to anything on telly."

## *Australian Cartoonist*

Michael Leunig, is an extraordinary cartoonist that had taken Australia by storm with a series of satirical pictures of almost everything and anyone in the 1960s. I was most impressed with what I still consider the best cartoon that anyone had ever done on their observation of everyday life. Imagine a fairly bare room with a television set, where a man and his young son sit watching a show on the television. Both the show and the window show the sun rising from behind the mountains. It made a beautiful picture.

It was sometime during the 1990s when I first came across Michael's cartoons. They appeared regularly in the daily papers and in his soft-cover books. Much later, when I was living on Bruny Island, he came down for a talk about his life to the locals. Marian and I attended. After the talk, we were quietly sat enjoying a cup of tea, and I asked him if he would be good enough to sign the cartoon book that I had brought with me. He happily obliged and we spent quite some time talking. Michael agreed with me that the best drawing he had done was the sunrise. I'm sorry to say my book has now disappeared on the move from Bruny Island to Queensland where I lost quite a lot of my important memorabilia.

## *Marian Passes Away*

In 2017, Marian had a stroke on a Monday morning. I had been working in my library and when 11am came around I shouted, *"Let's have a coffee!"* As I came around the corner, there she was, lying on the floor. I instantly phoned the doctor. Everything happened so quickly, it felt like only 15 minutes had passed by the time the doctor arrived and had her securely aboard a helicopter, to take her to the hospital. Her sister flew down that day and was quickly by her bedside. She stayed at the hospital to look after her. We both knew the end was near. Marian died four days later, on Friday and I sorrowfully travelled back home.

## *Inala Bird Sanctuary*

In the earlier years, long before Marian's death, we had entertained Dr Tonia Cochran, who had a sanctuary for the birds on Cloudy Bay Road, a few kilometres up the road from where we lived. Every New Year's Eve, Tonia would invite us to her spare block of land, being at the highest point on Bruny Island overlooking Cloudy Bay, to have a BBQ. We would all sit at her camp table set up with her large stone statue, maybe a metre high, of an eagle that her ex-husband, Colin, had put there. Even though they were divorced, they remained good friends. Colin came down and stayed in her holiday units one weekend and I remember when he left on the Monday, to go home, he'd left half a bottle of Grange Hermitage. Tonia asked if I would like it... *"Yes please,"* I replied. In 1970 Grange Hermitage was $18, I discovered in a book on Guide to Australian wines, but today would fetch about $900.00. It was a lovely wine, I enjoyed drinking it.

Tonia had a boyfriend, Scotty, who was the engineer on the Bruny Island ferry, *Mirambeena*. We saw him on every trip into town and on the way home, so we soon became very close friends. Sadly Scotty died in 2018 from complications but Tonia continued to come for a meal at my house, usually every Friday night. I took pains to prepare a good meal, mainly because she showed so much enjoyment, not only for being here but for the food as well. I still miss her and we keep in contact every so often.

It was my pleasure, when I left Tasmania in 2020, to donate my collection of old glass bottles for inclusion into Tonia's museum that she built in her bird sanctuary. To this day she still receives enormous numbers of tourists who come from all over the world to view the many plants she was growing that also grew during the Jurassic period in history. It was possible to view many rare forty-spotted pardalote birds, which she had living on her property. She had built a platform into the trees that one could climb up and watch the birds from the tree tops.

### *Jacquie Passes Away*

My first wife of fifteen years and the mother of my children, Jacquie, died on the 3rd June, 2018. We were married in 1956 and separated in 1972.

# Love of Music

Music has played a big part in my life. From the moment those first strains of the 'Marche Militare' by Franz Shubert entered my very young brain, I surrounded myself with music at every opportunity. My taste changed a great deal but I have never altered my love for the great classics. Popular music, jazz, country and western etc have come and gone, except for perhaps a few numbers I now always enjoy. Benny Goodman (1909-1986), Lionel Hampton (1908-2002) and Glen Miller (1904-1944), and Django Reinhardt (1910-1953) will always be my favourites but are no longer known on radio. Yes, it's radio that I am really concerned with, as it is always there; one takes what one gets.

I've spent a fortune on record players, recorder reel to-reel tape decks and hi-fi equipment. Over the years my music collection has included: thousands of 78s of both 10 and 12 inch records, and tapes and CDs of classical music. Today I have a mobile phone and can just dial up almost anything I want to hear.

It's the ABC that has been my main listening station from 2FC and 2BL on the days before FM came in. What a joy that was; the quality of classical music rose to great heights for enjoyable listening, nearly as good as a live concert, which never went astray in my quest for more music.

### *Sydney Town Hall*

When a teenager at school, the Sydney Town Hall offered free admission on a Sunday morning to listen to the rehearsals of the following week's artists. I had a music teacher, Mr Poole, who would accompany me to these free concerts. He taught mainly classical music but had a few followers, like myself at the school. At this time I had my original radio, a STC, in a wooden cabinet and I became interested in what all those tubes were doing when I looked in the back of the set. Years later I removed the back board of the cabinet and noticed it just had the bare chassis where one could observe the working parts.

It was a bit of a battle to choose between 2BL or 2FC. Margaret Throsby had a trivia program called Q&A. I got involved in this, trying to answer a lot of the questions. When the program had been running for a few years, she produced a number of books. At Sydney Town Hall the first issue was launched and I went along to see, not only to buy a copy but because I wanted to give Margaret a big kiss on the cheeks for being the most fabulous person, I enjoyed her show so much.

Over the years I made lots of friends from the ABC by phoning in with queries. When Guy Noble came down to Tasmania, I got up early and went to the station to meet him during his breakfast program, where he interviewed me and showed great delight in my effort to meet him.

My sister, Sally's, friend Emmanuel was a guitarist. Although I did not ever hear him play, I did hear his recordings after becoming good friends with his wife, Mandy, after he died. I met Mandy in Nambour, Queensland as she was a good friend of Sasha. We had many good times listening to his music.

Marian was a great fan of Richard Tauber (1891-1948) and had nearly all of his records, which we listened to on many occasions when we lived at Cloudy Bay. In fact, at this time, along with Tauber, we collected a lot of operatic singers. As my knowledge grew, I became a Luciano Pavarotti (1935-2007) fan.

# Special Friendships

What have friends meant to me? What is friendship? Do we take it for granted or do we think about it when we meet someone we like? No, it just happens, we don't plant it, we don't take it for granted - it just happens, like love. When found it can last a long time and to lose a friendship can be devastating and hard to get over. I have found the most profound friendships in my life are nearly always with another male. Friendships with women can and do exist but are much rarer in life. I've been very lucky in life to have many friends who are valued as close or very close. Some last a lifetime and some come and go like the weather.

### *David and Ian*

My finest experience at having a friend turned out to be a neighbour living in Mosman Bay. David and I were both about four years old, in 1935. Living just up the road from our homes was another boy of our age and his name was Ian. Our friendship is still alive as much then as it is today. We went to the same kindergarten, and joined the same scout movement and wolf cubs, but ended up being sent to different schools. We remain firm friends though getting together now is not so easy – but it does happen on occasion.

Ian, me and David reunion (2011).

## *George and Barney*

It was not until I moved back to Western Australia in 1942 that I formed another friendship with someone who was living not far from our house. What struck me the most about George was his dog. A very large wonderful Irish Red Setter that went by the name of 'Barney'. Except for school, when we both went to Wesley College, Barney was always with George. A couple of years followed and I begged my dad to get me an Irish Setter too. My dog turned out to be a she so I called her 'Pat'. Barney, Pat, George and I were inseparable. We swam, caught crabs and we fished for cat fish in the river.

We even built a small sailing boat together. One day we acquired a sheet of corrugated iron and while road works were being carried out, we carried the sheet over to the man driving a steam roller. We asked him to run over our sheet till it was almost flat. He obliged and we then carried it back to George's place with a little help from his brother. We folded the two ends together between a piece of wood and covered the areas as we assembled it with tar that we had collected off the road on hot days. We wrapped cloth around the edges to avoid getting cut on the tin and filled any little holes found in the tin with tar. We formed two sheets, using nails and timber to join them together. Then we joined the ends to the rim and fitted a broom handle with one of my mother's worn-out sheets for a square sail. We bravely set forth across Perth waters; not the most successful venture but we agreed we had a magnificent time.

In 1946 I returned to Sydney and didn't see George again until the day I was back in Perth on the ship bound for London. On my last trip to Perth, he had died just days before my arrival.

## *Ian*

As you've read through this book, Ian has been a major part of my life. I married twice and quite remarkably he managed four wives. He was best man at both of my weddings but I have yet to be asked to be his. Ian wrote a whole string of books that not only sold well but became Hollywood movies. We catch up every so often as we both get a little greyer, stooped and I admit it, a bit slower. I do wonder how many gallons (litres) of grog we have shared over the last fifty-some years.

Beatrice and Ian at my Wedding reception in London (1956).

## *Garth*

I had developed another friendship with a lad from the scouting days: Garth. In boxing training in the scouts, they would put us in the ring together. I thought, *"How can I hit a friend?"* And then I was hit by his friend, a knockout, and feeling quite embarrassed, that was the end of that.

Garth was hard to get moving at times; not like me, I just worked to have fun. But he buried himself in law books till at a much later date he became head of law at the University of NSW. He was a very clever man and taught many people the art of law. He met the Dalai Lama and did much work for the Timorese people until his retirement. I was living in Tasmania when I received a long letter from Garth asking if he could come and visit. He stayed with us for four days. Garth explained that during his career in law he had made very few friends. What a wonderful reunion we had. From then on, he and his girlfriend came down every other year to visit me in Tasmania. We had some great times together, until, sadly, he died. God I miss him.

### **Wayne and Bob**

Six-foot-tall Wayne Bottom was a member of the NSW Police

Force and his wife was Marian's close friend, Michelle Small. Wayne was the spokesman who announced any policing changes or crimes that had occurred in New South Wales.

Wayne's brother, Bob Bottom, was a well-known journalist, being a crime reporter and writer. He became close to various criminals, being trusted, which gave him vital information as to what was going on in the underworld. He crossed paths with the notorious gangster, Trimbole, but covered many other well-known crimes of the day. Bob was under 24-hour police protection for 11 years due to his many threats of death. When he came to visit us, we were living in Pitt Street. He would never arrive until 9.30pm. If we were having dinner, we always postponed it until 10pm so he could join us in the dining room.

When Wayne and Michelle moved down to Wollongong, we missed them immensely. Bob later passed away, and I lost a very good, reliable friend.

My closest friends from the 1st Mosman Rover Crew at Crows Nest in Sydney (2002). George Skinner, David Johnson, Ian Pilz, Garth Nettheim, me and Ian Hamilton

# CHAPTER 14

## Sunshine Coast (2020-2024)

I had been on my own, for nearly three years after Marian's passing when I thought I would come and join my daughter, Sasha and her husband Benno, up in Queensland for the warmer weather and to be near my family. I sold my castle at 884 Cloudy Bay Road, Bruny Island, to a multi-millionaire from Victoria. He bought the house, the land and all our antique furniture.

I knew I wouldn't need to buy a big house in Queensland, thinking I might rent a flat or unit. However, I did bring a container of my treasured tools and some books. Initially I moved in with my daughter. But soon after arriving in Nambour, I found a house that suited me perfectly in the nearby hinterland town of Flaxton. It was a three bedroom house, the third bedroom being small, perfect for an office space. A large living room, dining room and a big garage.

I joined Sasha and Benno on a Variety Bash, driving through the many outback towns, and stopping to meet the locals. The cars had to be at least 30 years old, it was a lot of fun.

Sasha and Benno each received their Master of Mental Health Nursing in 2010 and later formed a Suicide Prevention Group in Nambour after my granddaughter lost her partner to this terrible illness. I volunteered my time supporting this group and met some wonderful people.

Benno and Sasha (2010).

From January it rained heavily for a month. By February my house became flooded and I moved into Sasha and Benno's granny flat while repairs took place. During my time in the granny flat, I commenced writing this book. By the 10th July I was still at Sasha's place.

When I finally put down my pen on 27th March 2023, since the 31st October, 1931, I had lived through 33,689 days. That's 33689 sunrises and 33689 sunsets. To select from those 33,689 days, I can remember the many wonderful places, seasons, people and events, from a small fraction of a moment to a full day of unimaginable and unforgettable memories, cramming them into these few pages. I can slap a title and picture onto the front cover, a blurb on the back cover and now you have my entire lifetime of thoughtful recollections.

# Thirty-six abodes

## AUSTRALIA

| | |
|---|---|
| 1931-1935 | 5 Hardy Street, South Perth, WA. |
| | St Kilda Road, Victoria. |
| 1934-1936 | 1 Rose Crescent, Mosman, NSW. |
| 1936-1939 | 10 Mosman Street, Mosman, NSW. |
| 1939-1942 | 4 Park Avenue, Mosman, NSW. |
| 1942 | 1 Queen Street, South Perth, WA. |
| | North Sydney, NSW. |
| | Prince Alfred Road, Mosman, NSW. |
| | Great Western Highway, Blackheath, NSW. |
| | 120 Middle Head Road, Mosman, NSW. |

## ENGLAND

| | |
|---|---|
| 1955-1956 | Hotel Kensington, Kensington, London, England. |
| | Earls Court, London, England. |
| | 38 Birchington Road, Crouch End, London, England. |

## CANADA

1957-1961　　Hotel Montreal, Montreal, Canada.
　　　　　　　Ridelle Avenue, Toronto, Canada.
　　　　　　　Kitchener, Toronto, Canada.

## NEW ZEALAND

1961　　　　　Crum House, Auckland, New Zealand.

## AUSTRALIA

　　　　　　　Rose Cottage, Russell Avenue, Faulconbridge, NSW.
　　　　　　　Bulkara Road, Bellevue Hill, Sydney, NSW.
　　　　　　　120 Jersey Road, Woolhara, NSW. (owned)
1962-1970　　122 Hopetoun Avenue, Vaucluse, NSW. (owned)
1970-1971　　Philips House, Elizabeth Street, Redfern, NSW (workplace leased).
1971-1973　　Hotel Taree, Taree, NSW.
　　　　　　　House in Taree, Taree, NSW.
　　　　　　　Block of land, Jones Island, Taree, NSW. (owned)
1973-1975　　Paling Place, Beacon Hill, NSW (Ian Hamilton's place).

|  |  |
|---|---|
|  | 4 Wheeler Parade, Dee Why, NSW. |
| 1975 | 7/30 Cook Road, Moore Park, Sydney, NSW. |
| 1975 | 215 Cleveland Street, Redfern, NSW. (owned) |
| 1978 | 56 Pitt Street, Redfern, NSW. (owned) |
| 1996-1997 | 444 Main Road, Bruny Island, Tasmania. (owned) |
| 1996-2020 | 884 Cloudy Bay Road, Bruny Island, Tasmania. (owned) |
| 2020-2024 | 25 Nimbus Drive, Flaxton, Queensland. (owned) |

# Climbing the Bridge

One more story to tell... In 1949, at our regular meeting for the 1st Mosman Rover Crew, we were drinking coffee when someone suggested we go for a night out on the town in Sydney, have a few beers, take in a show, and then dine at a Chinese restaurant. We watched a good vaudevillian (live theatrical) show at the Old Tivoli, then went down to the city markets where the best Chinese eating houses were. The night was most enjoyable and after a good feed, being quite tiddly, we didn't want to get on a tram so we decided to walk down to Circular Quay in time to catch the last ferry back to Mosman.

Taking our time strolling along, looking in the shop windows as we made our way down to Circular Quay, we arrived to see the last ferry departing the jetty. The group agreed to walk home and climb the Sydney Harbour Bridge arch. No one in our group had ever climbed the bridge before; it was a first for all of us.

As we approached the pathway, we passed the point where the main arch cut off through the road and disappeared into the sky. There may have been a few second thoughts at that point. We climbed over the wire gate ignoring the sign "£5 fine for climbing on the arch". Looking the other way, we all climbed up. At that point, I had no recollection of the hour, but we pressed on. It was no easy feat being in total darkness but thankfully we had leather shoes on and the arch had little metal steps and a wire rope balustrade to guide us upwards.

Arriving at the top, 134 metres high, we all relieved ourselves into the Harbour below, sat down and waited for dawn. The sun rose magnificently and what a splendid sight it was from our high perch.

At this point we decided to leave, but someone noticed a policeman directly below looking up at us, frantically pointing to the southern end for our climb down. We indicated yes we would, however, we had no intention of following his orders and scampered down the northern end. Quickly clearing the gate below, we scurried down the stairs at the railway station, with no sign of the policeman. Thankfully we simply walked home.

Our escapade happened some seventy years ago and it is still fresh in my memory as if it happened yesterday. It was the wrong thing to do but it made some wonderful memories for all of us.

Sydney Harbour Bridge sunset and me (2005).

# EPILOGUE

Terry passed over peacefully on Monday 17th June 2024. His ashes now rest beneath the arches of Sydney Harbour Bridge in the very space where his spirit found both solace and inspiration. It is here beneath the grandeur of the bridge that his journey finds its serene conclusion blending seamlessly into the narrative of a city he admired so deeply.

Though Terry has departed from this world, his presence remains vividly etched in the hearts and minds of those who knew him. His legacy endures in the stories he shared, the wisdom he imparted and the love he radiated. As readers close this book they carry forward the essence of Terry's spirit enriched by the reflections of a life well-lived.

Terry's life was a testament to resilience and adventure with a deep-seated love for the world around him.

# ACKNOWLEDGEMENT

Sasha and Benno would like to express their heartfelt thanks to the following people:

Rolf, Bev, Alexa, Ellie, Holly and Tess

Ayla, Bronte and Cleo

Sally, Jane, Libby and their families

Erica and Jane

Tonia Cochran of Inala Nature Tours, Bruny Island

Claudia Shaw and Stewart White

Members of the Bruny Island Mens Shed

Members of the Tool Association Sydney including Ray and Mike

Members of the Suicide Prevention Group

Dave Johnson, his lifelong mate from the 1st Mosman Scouts Club

Paul Keating, former Prime Minister

Kate and Bruce Ganley

Ellie and John McPherson

Vivienne and Garth Mackie

Hans Brunner, antique dealer friend

State Library of Western Australia

"Terry made considered and generous donations of important heritage records from the Butcher family collection that provide a legacy for the State. I am pleased that preserving and sharing these treasures through the State Library ensures that they are available to be discovered by historians, researchers and other community members here in the Library or in digital format from anywhere in the world." State Library of Western Australia (31 July 2024)

# BIBLIOGRAPHY

- Barnes Wallis: A Biography (1972) by J E Morpurgo.
- Burchell's Travels, The Life, Art and Journeys of William John Burchell 1781-1863 (2015) by Susan Buchanan.
- Heroes Galore (1996) by Richard Timperley.
- Jewish Women Jewish Men: the legacy of patriarchy in Jewish life (1995) by Aviva Cantor.
- Montsalvat, the intimate story of an Australian artists' colony (2014) by Sigmund Jorgensen.
- The Road to Bulong, "Close up – a little bit long way" (2007). A history of the Jones family of Hamilton Hill by Pamela Rajkowski and Elizabeth Tuettemann.
- Ups and Downs, Pioneers of Land and Air (2001) by Wendye E. Camier
- Sparks of Genius, The inventive photography of Bill Angove (2024) by Richard Goodwin.

www.ingramcontent.com/pod-product-compliance
Lightning Source LLC
Chambersburg PA
CBHW062030290426
44109CB00026B/2582